UITGAVEN „NATUURWETENSCHAPPELIJKE STUDIEKRING VOOR SURINAME EN DE NEDERLANDSE ANTILLEN"

PUBLICATIONS OF THE FOUNDATION FOR SCIENTIFIC RESEARCH
IN SURINAM AND THE NETHERLANDS ANTILLES

Obtainable at the Secretariat:

Zoological Laboratory, Plompetorengracht 9–11, Utrecht, Holland

STUDIES ON THE FAUNA OF SURINAME
AND OTHER GUYANAS
XV

STUDIES ON THE
FAUNA OF SURINAME AND
OTHER GUYANAS

EDITED BY

Dr. D. C. GEIJSKES
AND
Dr. P. WAGENAAR HUMMELINCK

VOLUME XV

With 47 text-illustrations and 18 plates

Uitgaven „Natuurwetenschappelijke Studiekring voor Suriname
en de Nederlandse Antillen", No. 80

SPRINGER-SCIENCE+BUSINESS MEDIA, B.V.
1975

ISBN 978-94-017-7079-8   ISBN 978-94-017-7106-1 (eBook)
DOI 10.1007/978-94-017-7106-1

CONTENTS

STUDIES ON THE FAUNA OF SURINAME AND OTHER
GUYANAS: No. 56

HYDROBIOLOGICAL OBSERVATIONS IN SURINAM

WITH SPECIAL REFERENCE TO THE MAN-MADE BROKOPONDO LAKE

by

P. LEENTVAAR

(Rijksinstituut voor Natuurbeheer, Leersum, Nederland)

CONTENTS

2

Address of author: Research Institute for Nature Management (R.I.N.), Kasteel Broekhuizen, Leersum, The Netherlands.

1.   INTRODUCTION

The construction of large reservoirs such as the man-made Brokopondo lake, is certainly not the result of proposals and conclusions of biological studies, but rather of political, technological and economical decisions without serious consideration of the biological implications. The biologist is faced with the results of a serious and hazardous intervention in the environment of man, animals and plants, which must be evaluated and if possible managed after planning and construction. Therefore it is a positive development that in recent years in more cases the ecological aspect of dam construction is integrated in the plans.

In the case of Lake Brokopondo, officially called Prof. Dr. Ir. W. J. van Blommesteinmeer, long before the work was started the biological implication of the construction of a dam in the Suriname River was considered in a scientific study (SCHULZ 1954) sponsored by the Foundation for Scientific Research in Surinam and the Netherlands Antilles (STUDIEKRING), which study however, appeared to have no influence on the planning. The "barrage" was built by the Suriname Aluminium Company (SURALCO) for hydro-electric purposes and completed on February 1, 1964. In 1962 the Executive Board of the Netherlands Foundation for the Advancement of Research in Surinam and the Netherlands Antilles (WOSUNA) allocated funds for the carrying out of the Brokopondo Research Project and so in November 1963 the hydrobiological investigations in the future lake region could be started. Some months later also botanical and ichthyological research began.

It goes without saying that this was for me an unique oppoi tunity to study hydro-biological phenomena in the unknown tropical inland waters of Surinam and to follow the forming of one of the biggest tropical impoundments about 1500 square kilometres, comparable only with the well-known barrage lakes Kariba, Volta, Kainji and Aswan in tropical Africa. I am most grateful for this to the Board of WOSUNA and the director of the State Institute for Nature Management (Prof. Dr. M. F. MÖRZER BRUYNS) who enabled me to undertake this study. My experience with the hydrobiological work in the Netherlands, focussed on evaluation, conserva-tion and management of waters for nature conservation, could now be used in this tropical country.

Arriving in Surinam a few months before the closing of the dam, I immediately started to collect as much data as possible about the still flowing Suriname River. Without the help of SURALCO, who provided boats, manpower, laboratory, housing and other facilities at Afobaka this would not have been possible. Also the assistance of the Government through the District Commissioner of Brokopondo was of great value in providing permanent manpower and in arranging the expedition on the Saramacca River in April 1964. Mention must also be made of the kind assistance of Dr. D. HEINEMANN from SUNEVO (Surinam Netherlands Health Organisation) at Paramaribo, who made it possible to carry out more extensive chemical analyses in his laboratory, which could not be done in the primitive hut at Afobaka.

The Biological Brokopondo Research Project was set for a period of three years. During this time the team of biologists was backed by the Netherlands Commission for Research in Artificial Lakes in Surinam of the STUDIEKRING. In 1964 the WOSUNA was reorganized in the Netherlands Foundation for the Advancement of Tropical Research (WOTRO), which took over all financial and administrative responsibilities. The assistance of the "Brokopondo Commission" of the STUDIEKRING, and especially of its president Dr. J. H. WESTER-MANN, may be considered to be vital for the work and the project. In Surinam the social worries of the team were ably solved by the representatives of resp. WOSUNA and WOTRO, which saved much time. It will be understood that it is impossible to mention all people who were helpful during the stay in Surinam.

Some months after my start of the hydrobiological research at Afobaka, I was joined by Drs. J. VAN DER HEIDE, who continued the regular observations when I left Surinam in 1964. In 1967 he was succeeded by Mrs. drs. H. NIJSSEN-MEYER, who finally closed the continuity of the work by leaving Surinam in July 1967.

The STUDIEKRING and WOTRO decided to continue the research only by giving facilities to investigators for short visits. However much we appreciate this gesture, it must be stated that regular continuation of the hydrobiological observations in this newly formed tropical impoundment should have been taken over by some Surinam organisation for scientific research.

The Division of Hydraulics of the Ministry of Public Works and Traffic at Paramaribo (WLA) in 1967 started a monthly sampling programme at the fixed stations by measuring oxygen, temperature, BOD and other factors. The execution of this programme became possible as in August 1967 the lake was almost ready for operation. It was agreed that at this point the Government should take over some of the tasks for the management and operation of the lake from SURALCO. It was arranged that WLA included monthly plankton samplings in its programme which were forwarded to the Netherlands together with the chemical data for investigation and criticism. In this connection my visits to Lake Brokopondo in 1968 and 1972 must be considered and the results are added in the Appendices of this paper. In the years after 1967, WLA did not stop at this sampling programme but started more activities on the lake such as measuring evaporation and precipitation at different stations; measuring wind effect on water movements in a hydro-meteorological scheme. These observations already showed their value when climatological research was done by airborne remote sensing. So thanks to the activities of WLA we are able to follow the developments in the lake and we hope that this will be stimulated also from outside the country.

The international scientific interest in the Brokopondo Project was evident, especially from the limnologists working on other tropical reservoirs. In the preparatory phase of the research Dr. E. B. WORTHINGTON from the Nature Conservancy in London was invited to the Netherlands in order to discuss the programme with the team and also Dr. ROSEMARY McCONNELL née Low provided suggestions in relation to her experience with ecological research done in Guyana.

In the course of time since the beginning of the work at Brokopondo the contacts with other investigators were intensified. In most cases the interest went out to the growth of water-hyacinth, the rate of decay of the drowned forest and the problems with migration of indigenous people in the lake region. The interest was also stimulated by several lectures and publications and the regular distribution of the Progress Reports; in 1965 at the Man-made lakes Symposium of the Institute of Biology in London; a lecture in Plön at the Max-Planck-Institut für Limnologie, Abteilung Tropenökologie where the contact with Prof. Dr. H. SIOLI became close; in 1968 lectures for the International Society of theoretical and applied Limnology in Warsawa and in 1971 in Leningrad; a contribution to the Symposio sôbre a Biota Amazonia in Manaus in 1967; a lecture at the International Symposium on Man-made lakes at Knoxville, U.S.A. in 1971.

International interest was shown also by the assistance of several specialists for the identification of the various taxonomic groups. Dr. G. MARLIER, Brussel, identified a number of Trichoptera; Dr. G. DEMOULIN, Brussel, Ephemerida; Dr. F. WIEBACH, W. Germany, Bryozoa; Dr. Br. BERZINS, Sweden, Rotifera and Dr. P. BOURRELY, Paris, Desmidiaceae. In the Netherlands assistance was given by L. J. M. BUTOT from my institute (R.I.N.), Mollusca; Dr. L. B. HOLTHUIS from the Rijksmuseum van Natuurlijke Historie at Leiden, Crustacea. Dr. P. VAN DOESBURG, Leiden, Rhynchota and Coleoptera. Dr. D. C. GEIJSKES, Leiden, identified the Odonata and was also helpful with the identification of several other groups of organisms.

6

Mr. A. VAN DER WERFF, De Hoef, identified several of the diatoms. I am most grateful for their kind help.
Thanks also to Mr. H. WERMENBOL, R.I.N., who made the drawings (Figs. 1, 3–39).

Finally I must thank Mr. and Mrs. E. BLOK VAN CRONESTEIN, Zeist, who co-operated in the correction of the very first draft of the English text, and not in the least Dr. P. WAGENAAR HUMMELINCK and Drs. LOUISE J. VAN DER STEEN of the Zoological Laboratory, Utrecht, who with never failing criticism and lasting enthusiasm prepared this treatise for the press.

Methods

At different depths water samples were collected with a Ruttner water sampler of 1 liter volume.

Oxygen was estimated by titration according to the Winkler method. In 1968 oxygen was also measured with membrane electrodes, but lack of accuracy and difficulties with the instrument forced us to return to Winkler titrations.

Electric conductivity was measured with a Dionic meter; pH with an electric pH meter (Electrofact).

Plankton was collected by sieving through a net with 46 μ meshes. For quantitative purposes 4 buckets of water (about 40 l) from the surface were poured through the net.

The plankton organisms were identified with the aid of several keys, such as those found in the well-known volumes of THIENEMANN's *Die Binnengewässer* and *Die Tropische Binnengewässer*, and further in: HUSTEDT's *Die Kieselalgen*, in RABENHORST's *Kryptogamenflora*; PRESCOTT, *Algae of the Western Great Lakes Area*; GRÖNBLAD, *De Algis Brasiliensibus*; SCOTT & GRÖNBLAD, *New and interesting Desmids from the Southeastern United States*; VOIGT, *Rotatoria*; WARD & WHIPPLE, *Fresh-water biology*.

The results of the examination of the plankton samples are given in Tables I–XIV. An estimation of the relative abundance of each species is given as follows: 1 = present, 2 = moderate numbers, 3 = many specimens, 4 = large numbers, 5 = very large numbers.

2.   RIVERS AND STREAMS

Most rivers in Surinam run from south to north into the Atlantic Ocean. In the east the large Marowijne River (Maroni) borders French Guiana; in the west the Corantijn River borders Guyana (formerly British Guiana). These two are the largest rivers. In between, smaller rivers are found e.g. the Suriname, Saramacca, Coppename and Nickerie. In the coastal plain a number of small rivers originate at only a short distance from the coast e.g. the Boven Commewijne, the Coesewijne and Tibiti, the Maratakka and Nannie Kreek. Some of these rivers run through the sandy soils of savannas, from which they derive a characteristic brown colour, so that they may be characterized as Brown Water. In the tidal zone close to the coast some rivers run in an east-west direction, as for instance the Cottica, connecting the Marowijne and Suriname River.

2a. SURINAME RIVER (Figs. 1–3; Table 1)

The Suriname River flows between lats. 6 and 3 N and longs. 55 and 56 W to the Atlantic Ocean. The drainage area is 12.200 square kilometers. Near the sea a flat coastal plain is found and from there the land slopes upwards to the watershed. In the interior the subsoil is rocky; the topsoil is poor in minerals. For the greater part this area is covered with tropical rainforest. Sandy savanna areas are found in the low lying grounds, beginning at a distance of 70 km

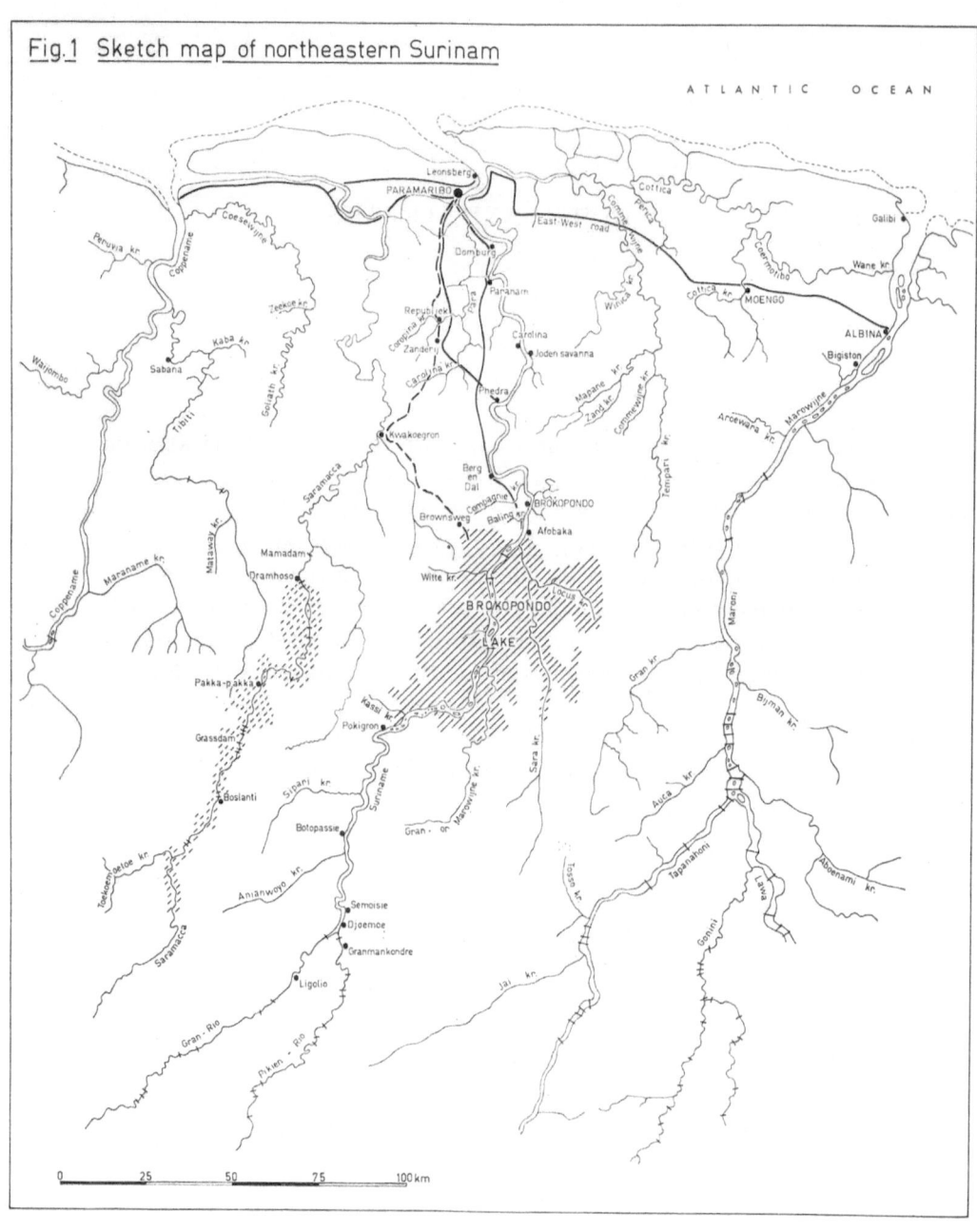

Fig.1 Sketch map of northeastern Surinam

from the sea. In the upper course many rapids and falls occur. Between these obstructions the river is sluggish. In the lower course no rapids are found and the water gradually becomes turbid, especially in the tidal zone.

A few months before the closing of the dam, I arrived in Surinam and as no hydrobiological information was known of the Suriname River, there was little time left to collect data of the river in its original condition. The dam is located at a distance of about 110 km from Paramaribo.

I began collecting data in a stretch of the river between Afobaka and Pokigron, where the reservoir would be formed. This barrage lake will eventually cover an area of about 1500 square km. Parts of the tributaries Sara Kreek and Grankreek or Marowijne Kreek will also disappear under water. Therefore these rivers were studied first. The results are recorded in the following chapters.

During the few trips I made on the river before the dam was closed, the only waterplants present were Podostemaceae, which occurred on the rocks of the rapids and falls. Along the borders no swamp vegetation could be found. The trees grew right down to the water. There were very few pools or swamps with stagnant water along the river or (probably) in the interior of the forest. A few specimens of the waterhyacinth (*Eichhornia crassipes*) were found. Upstream in the Sara Kreek near Soekroewatra there was a small swamp encroached upon by the waterfern *Ceratopteris pteridioides*, which was badly damaged by caterpillars.

The paucity of swamp vegetation and *Eichhornia* may be due to the swift current and the changes in waterlevel. During the rainy season the waterlevel rises swiftly and large areas of the forest are flooded. In the dry season the river is more than 300 m wide with shallow places and sandflats. The rocks in the rapids are mainly dry at this time.

Observations on the Suriname River outside the reservoir region were carried out at a later date. In August and September 1964 expeditions were made to the upper course of the river, where the Pikien Rio and the Gran Rio join the Suriname River. In the lower course of the river observations were carried out which, however, will be discussed in another publication.

Fig. 2   LAKE   BROKOPONDO   max. extension

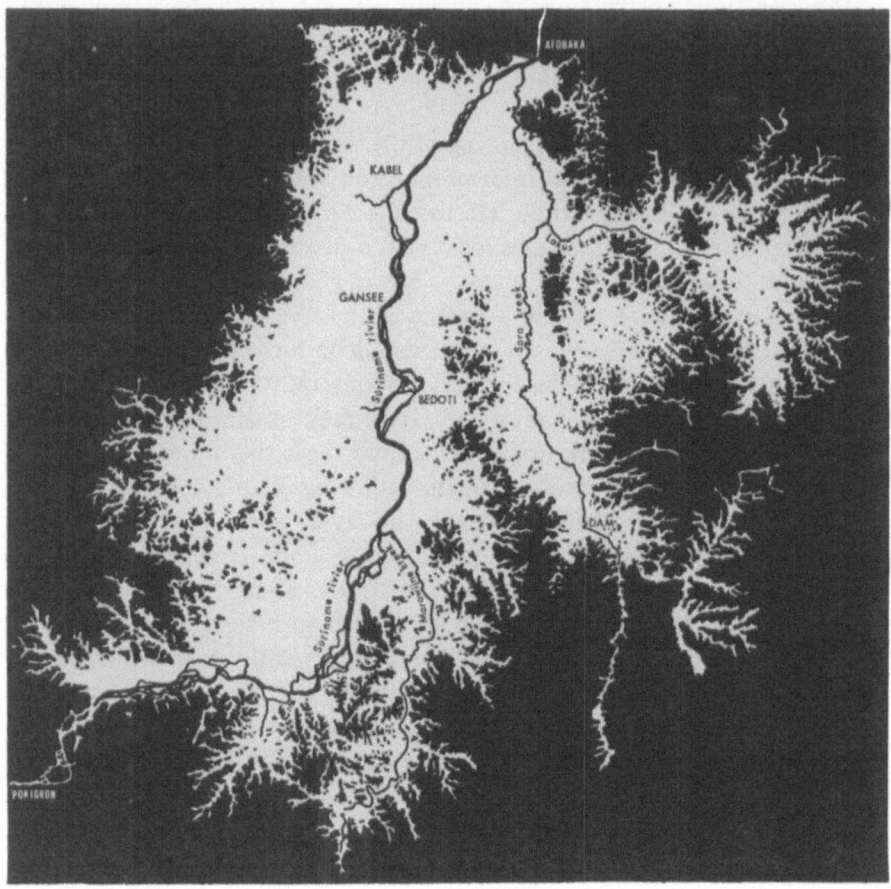

The upper course of the Suriname River did not differ much from
the stretch between Afobaka and Pokigron. Rapids with Podostemaceae and turbulent water alternate with slowly moving or even
stagnant water with sandflats. The bottom of the riverbed is rocky;
gravel and stones are absent. During the trips bottom organisms,
plankton and water samples were collected. The organisms found
are discussed in the next chapters. Data on the chemistry of the

Suriname River from the upper course down to the tidal zone are
given in Table 1.

The figures show that the Suriname River is very poor in minerals
and that the water is acid. During the long voyage to the sea the
mineral content gradually increases together with the pH. Data on
temperature are too few to justify conclusions, but it may be mentioned that other observations indicate that temperature rises as
well.

In the lower course, from Afobaka to Phedra, the river never
showed an electric conductivity higher that 35 μS. This was measured at Phedra, in the fresh water tidal zone.

The water of the river was only slightly coloured. From the air the
water looked brown, but in the laboratory, after sedimentation, it
appeared to be colourless. There was some opalescence due to iron
and silica; under field conditions, however, the small amount of
brown detritus and iron produced a brownish colour. There appeared
to be virtually no brown coloration due to dissolved humic acids.
The Suriname River therefore may be characterized as a Turbid-
Brown (iron) Water. This does not correspond with the water types
distinguished by SIOLI (1950, 1956, 1964) for the Amazon area. The
Brown or Black Water type of SIOLI comprises waters which are
coloured brown or even black by dissolved humic matter. The Suriname River has more affinity with the Clear Water type of SIOLI.
Black or Brown Waters are in fact present in Surinam but not in the
upper course of the Suriname River. Only in the tidal zone, some
Black Waters enter the Suriname River (Para Kreek, Coropina
Kreek, see also Chapter 2g).

The water is not very transparent. Secchi-disc readings were
1.25 m at Pokigron and 1.30–2.0 m at Afobaka.

Regular hydrobiological observations at Afobaka and in Sara
Kreek were only possible before the dam was closed (on February 1,
1964).

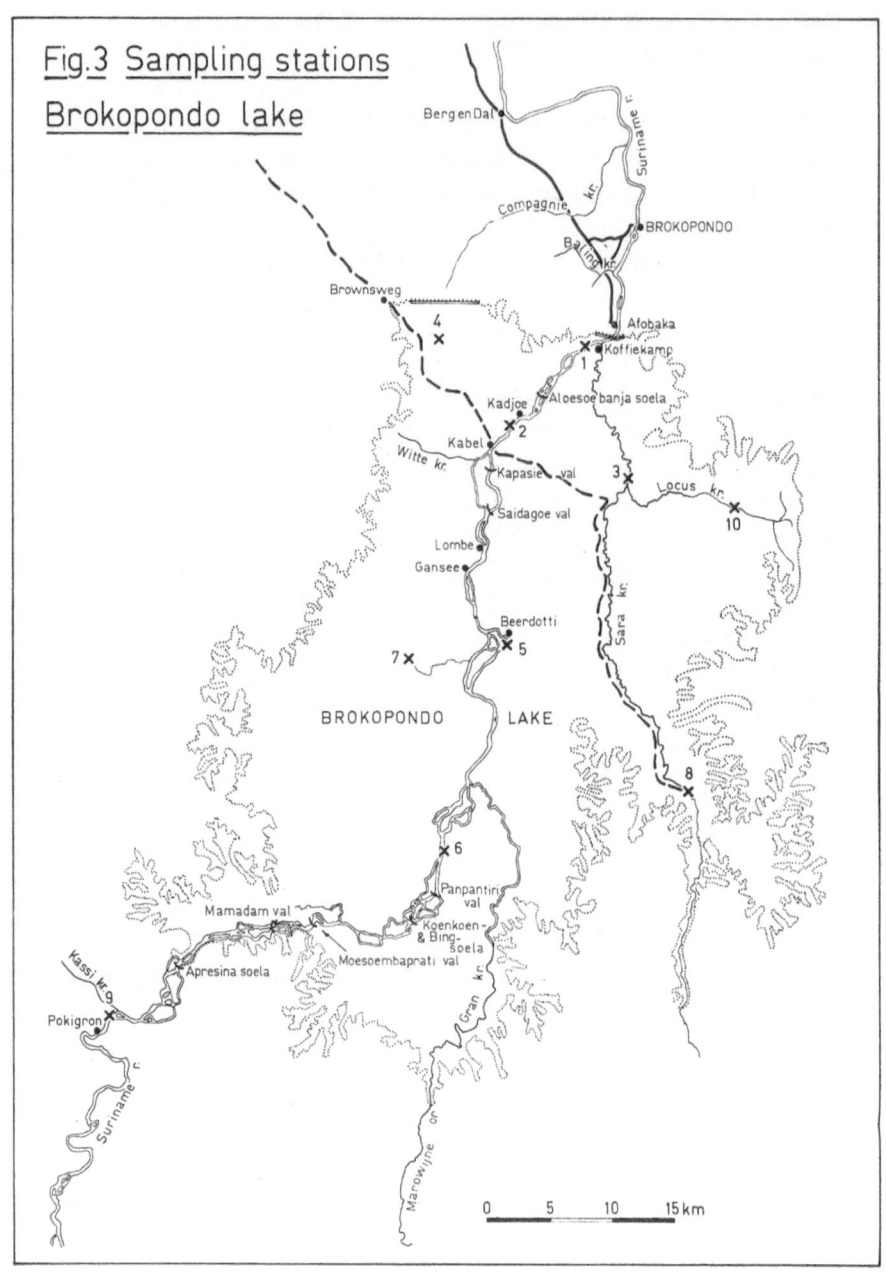

Fig.3 Sampling stations
Brokopondo lake

2b. SURINAME RIVER AT POKIGRON (Figs. 3 and 5–6; Table I)

As travelling by river from Afobaka to Pokigron took some days, it was impossible to collect weekly samples at this location. The Surinam Aluminium Company, however, took daily gauge readings at Pokigron and was so kind as to forward weekly samples to Afobaka. River gauges, temperature, electric conductivity and pH were recorded. Furthermore, two plankton samples were taken, one by pulling the plankton net some distance through the surface layers of the water and a second by pouring 4 buckets of water (40 l) from the surface through the net.

As the river at Pokigron will remain undisturbed after the lake has filled, at this station data about quality and quantity of the water entering the lake could be collected continuously. The observations from December 1963 to September 1964 give a good picture of the river during the course of the year. With certain restrictions they may serve to give an idea of the situation at Afobaka before the river was dammed. Observations from September onwards will be treated in a paper by J. VAN DER HEIDE (1974).

Fig. 5 shows that the waterlevel was fairly constant during the period from December until March. In the week of January 14, a short rainy period caused a rise. This also occurred in the week of March 25. In the last week of May the long rainy season began and the waterlevel was 2 m higher for a period of two months (June and July). Later on in August and September the waterlevel was lower.

The temperature during the observation period fluctuated from 25 to 31.3°C. If we compare the values found in the period from December ot February at Afobaka with those found in the Sara Kreek (Fig. 10 and 14) we observe that the temperature is more constant at Afobaka. The minimum temperature at Afobaka was always higher than that at Pokigron and in the Sara Kreek.

The difference is caused by the fact that in the Suriname River the water is warmed on the stretch from Pokigron to Afobaka. GEIJSKES (1942) has made the same observation going downstream on the Marowijne River. In the narrow and shady Sara Kreek the temperature was about 2°C lower than in the wide and exposed Suriname River at Afobaka. At Afobaka the water is warmest during

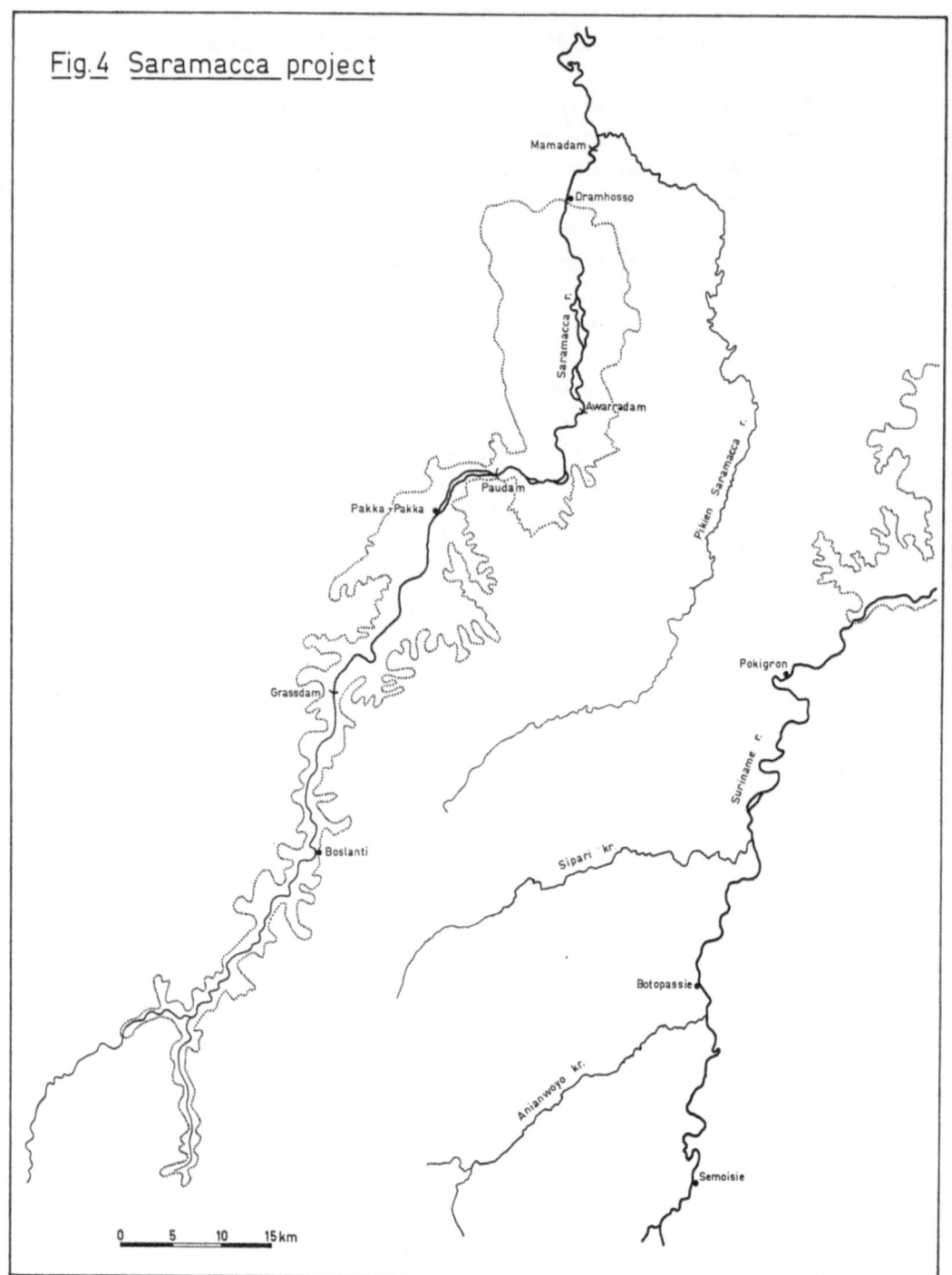

Fig.4 Saramacca project

CHEMICAL ANALYSES AND WATER LEVEL OF THE SURINAME RIVER AT POKIGRON (1963-1964)   5

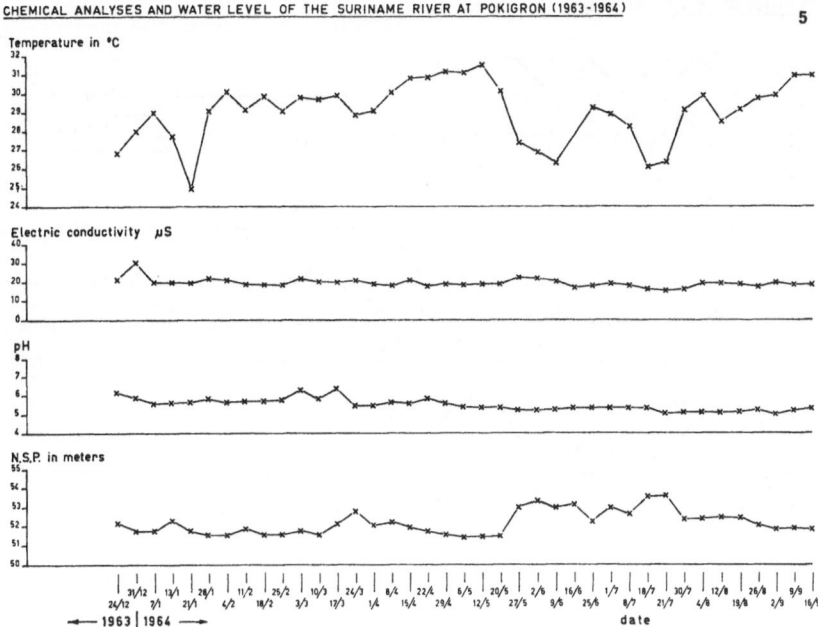

the dry period from April to May as the river then was shallow and sluggish. At the end of May, with the onset of the rainy season, the water rises and the temperature drops sharply. This picture is influenced, however, by events happening upstream of the sampling station. When local showers in an upstream region add cold water to the river it takes some time before it reaches the sampling station. As a result a drop in the temperature may be registered at a time when it is not raining anywhere near the station.

The electric conductivity is fairly constant at Pokigron, where it is lower than at Afobaka. The lowest conductivity (15 µS) is found during the rainy period at the end of July. However, there is a temporary rise in conductivity when the rains begin, due to the transport of material from the forest and river bed. The pH also is lowest in the rainy period. After some time all minerals, detritus and plankton have been washed downstream and from July 8 until August 26, nothing could be found in plankton nets apart from spiculae of sponges and a small amount of detritus. The water could

PLANKTON OF THE SURINAME RIVER AT POKIGRON (1963-1964)   6

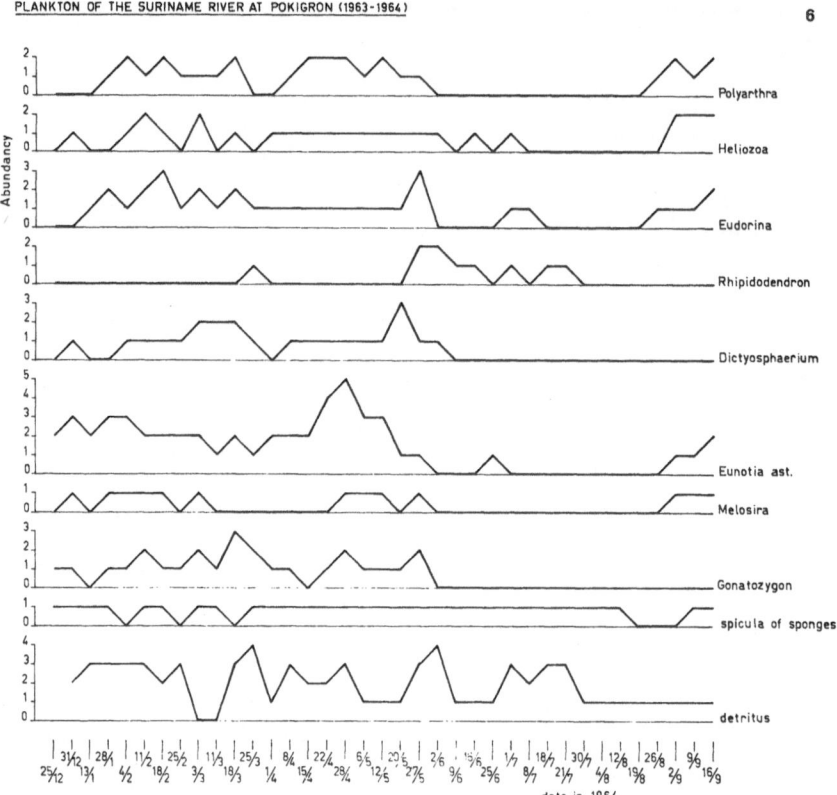

be considered to be practically pure rainwater. A characteristic
plankton community develops during the dry season when the river
runs slowly and is partly stagnant. When the rains start, the greatest
part of this community is washed downstream within a week. It is
replaced temporarily by a community composed of *Arcella*, *Heliozoa*,
a few *Eudorina elegans* and *Rhipidodendron huxleyi*, bottom diatoms,
mosquito larvae, Ephemerids and spiculae of sponges. This alloch-
tonous plankton originates from small rivulets and backwaters in
the forest. The general picture during the rainy period is that at first
river plankton is flushed and after that the plankton from the small
streams is transported. Finally only rainwater remains.

The plankton community of the river is very poor in zooplankton. In the dry season, April and May, we find the crustaceans *Cyclops, Bosminopsis deitersi* and *Moina,* and the rotifers *Polyarthra, Keratella americana* and *Conochiloides coenobasis* more frequently as the river flows more slowly then. Rhizopoda and Heliozoa are common in the plankton.

The motile green organism *Eudorina elegans* is present regularly and appears to increase in numbers when conditions in the water are changing, for example at the beginning of the rainy season. At this time more oxygen may be produced by the green organisms (*Eudorina*) and the river shows biological selfpurification under natural circumstances. GESSNER (1965) observed the same in the Orinoco. This phenomenon was also observed during the initial stages of the lake forming near Afobaka. The predominant non-motile organism in the plankton of the river was the diatom *Eunotia asterionelloides,* which developed in great numbers in the dry season. *Rhizosolenia eriensis* and *Rh. longiseta* were also common, *Surirella* occurred regularly and *Melosira granulata* was present irregularly and in very small numbers.

In the Saramacca river, which runs parallel to the Suriname River, *Melosira* is present in greater numbers and *Eunotia* is subdominant, while the total plankton community is similar to that of the Suriname River.

The desmids in the Suriname River are well represented by *Closterium, Cosmarium, Cosmocladium, Euastrum, Gonatozygon monotaenium, Micrasterias arcuatus, M. brasiliensis, Sphaerozosma granulatum, Staurastrum inaequale, St. trifidum, St. mamillatum,* and others. Chlorophyceae are less common. *Dictyosphaerium pulchellum* was found regularly, originating from stagnant parts of the river. *Pediastrum duplex* was seldom found in the river but it occurred in fair numbers in a grass swamp near Afobaka, and during periods of high water it could be transported into the river.

Blue-green algae were rare in the plankton. *Merismopedia convoluta* was found more frequently than *Oscillatoria.* Other organisms present in the plankton were nematodes, ephemerid larvae, mosquito larvae, Hydracarinae, zoecia of *Plumatella,* single eggs of dragonflies, and veliger larvae. The latter are free swimming larvae of a unionid, probably *Diplodon voltzi,* the most common mussel in the river. Sponge spicules and pieces of Podostemaceae in the plankton originate from organisms living on the rocks of the rapids. Detritus is always present in the plankton as a brownish black flaky material; the amount fluctuates as already mentioned.

The presence of *Actinocyclus normanni* was rather a surprise as this diatom is characteristic of freshwater tidal zones. Generally it prefers turbulent fast moving tidal waters, but there are no tidal currents in the upper reaches of the Surinam River. However, on a trip from August 13–17, *Actinocyclus* was found in the Pikien Rio, a Gran Rio tributary, the Gran Rio, the Suriname River at Semoisie and also in the Anjanwoyo Kreek. At the time of the trip the water was high because of the rains. Thus *Actinocyclus* is not restricted to conditions found in the tidal zones. As it was found in the upper reaches when rain caused turbulent water, it is probable that it occurred there as a tychoplankton organism, normally living on the bottom. The high specific gravity of the species causes *Actinocyclus* to sink in quiet water. The salinity of the water seems to be unimportant.

Apart from the regular observations a few incidental observations can be given. On December 24 and December 25, 1963 oxygen was 6.3 and 6.0 mg/l respectively, appr. 80% saturation. Similar values were found at Afobaka. The alkalinity was 0.20.

Transparency measured with the Secchi-disc was 1.25 resp. 1.25 m. This is somewhat less than at Afobaka. The depth of the river at the sampling place was between 1.25 and 2 m. All weekly samplings were taken at noon.

2c. GRANKREEK (Table 2)

The Grankreek (= Marowijne Kreek) resembles the Sara Kreek. It is a strongly meandering tributary of the Suriname River and many parts of it are overshadowed by trees. In the upper reaches many rapids are found and in the dry season navigation is not a simple matter.

In July a trip was made with dr. D. C. GEIJSKES and dr. M. BOESEMAN in order to get an impression of the part of the stream which would become part of the future Brokopondo Lake. Earlier observations were carried out a short distance from where the stream joins the Suriname River. The results of the chemical analyses are given in Table 2.

As in the Sara Kreek (see Chapter 4b), the electric conductivity in the Grankreek is higher than in the Suriname River. The pH is nearly the same. In July the conductivity and the pH were lower than in March and December, owing to the addition of much rainwater from the end of May onwards. The temperature of the water is lower than in the Suriname River as may be expected from the shaded character of the stream.

In December a large amount of brown flaky material with many iron bacteria (cf. *Leptothrix*) was present in the plankton, together with: *Alona, Cyclops, Anuraeopsis fissa, Polyarthra, Trichocerca similis, Arcella, Difflugia, Heliozoa, Lesquereusia, Eudorina elegans, Euglena, Phacus pleuronectes, Trachelomonas caudata, Dictyosphaerium pulchellum, Franceia?, Closterium, Cosmarium, Micrasterias arcuatus, Diatoma, Eunotia asterionelloides, Melosira, Surirella, Actinospora*, nematodes and single eggs of dragonflies. This plankton is like that of the Sara Kreek.

In July 500 m upstream only a small amount of flaky material was found, but many iron bacteria were present. Other organisms were *Alona, Bosminopis deiteris, Conochiloides coenobasis, Arcella, Eudorina elegans, Closterium, Actinocyclus normanni*.

Near the first and second rapid much brown flaky material was present in the plankton, together with Harpacticidae, *Testudinella mucronata*, many iron bacteria, *Arcella, Simulium, Eudorina elegans, Spirogyra, Actinocyclus normanni, Surirella* and *Lyngbya*.

The rapids were overgrown with Podostemaceae among which several organisms were found, notably *Doryssa devians, Pomacea granulosa, Pomacea sinamarina*, ephemerids, the freshwater shrimp *Macrobrachium olfersii* and crabs.

2d. SARAMACCA RIVER (Fig. 4; Tables 3 and II)

In view of the plans to create another artificial lake in the Saramacca River and for comparison with conditions in the Suriname River, a trip was made from Kwakoegron to Boslanti with dr. M. BOESEMAN from April 3–10, 1964. This trip was facilitated by the kind assistance of the District Commissioner of Brokopondo and the Police Commissioner of Saramacca.

The Saramacca River flows west of the Suriname River. It has a smaller catchment area, and in the dry season it is not easy to navigate because of soelas (rapids) in its upper course. A tributary, the Pikien Saramacca, has its source between the Saramacca and the Suriname River.

Samples of plankton and water samples were taken at various locations. The results are given in Table II.

The water in the upper reaches of the Saramacca River was browner than that of the Suriname River between Pokigron and Afobaka but this colour is also caused by brown particles. The river therefore may be typified as a Turbid Brown (iron) Water. Below Mamadam a large number of sandbanks are found and the river flows along the savanna. Near Kwakoegron a thin layer of silt was observed on the sand and the water was rather muddy and greenish. This is the head of the freshwater tidal zone.

The pH and the electric conductivity increase downstream. Compared to the Pokigron-Afobaka stretch of the Suriname River these values are lower. The water contains less electrolytes and the pH is appreciably lower than in corresponding sections of the Suriname River, particularly in the lower reaches near Kwakoegron.

This is probably due to the character of the Saramacca River which has many rapids and fast flowing water over longer stretches. Also the total discharge of the river is smaller.

The composition of the plankton of the Saramacca River was similar to that of the Suriname River, except that the dominant diatom *Eunotia asterionelloides* of the Suriname River occurs less frequently and is replaced by *Melosira granulata*, which occurs in both rivers.

In the upper reaches of the Saramacca *Nitzschia* and *Surirella* were frequently found. In the rapids and downstream to Mamadam *Melosira*, *Dictyosphaerium* and *Staurastrum* dominated. Further downstream in a slower moving section large quantities of filamentous *Gonatozygon* were present and also other desmids like *Closterium*, *Cosmarium* and *Micrasterias mahabuleschwarensis*. Finally, at Kwakoegron an increasing amount of green algae was observed whereas organisms of the types found in the faster flowing upper reaches were mostly absent.

It may be concluded that different communities exist in the river, resulting from the differences in the rate of flow in the various sections. Occasionally there may be some influence of water from tributaries joining the Saramacca, but it was not noticeable during this trip. The Pikien Saramacca did not carry much water and did not even communicate with the Saramacca. The samples were taken from a pool of stagnant water between the sandbanks near its mouth. The electric conductivity and pH were higher, as may be expected. The single plankton sample taken from this pool is not representative of the plankton composition of this river. It is none the less worth mentioning that the composition was roughly similar to that in corresponding pools in the Suriname River during the dry season. The similarity is demonstrated by the predominance of *Eunotia asterionelloides*, the paucity of *Melosira* and the lack of species which were generally present in the Saramacca. Consequently the Pikien Saramacca appears to have more affinity with the Suriname River than with the Saramacca, which may be explained perhaps by its location near the catchment area of the Suriname River.

As in the Suriname River *Eichhornia* was seldom found in the Saramacca. A striking feature of the rapids above Mamadam was the heavy growth of Podostemaceae. The rocks of the Awarradam were thickly overgrown with freshwater sponges. As in the Suriname River, water snails such as *Doryssa* and *Pomacea* had severely corroded shells due to the low pH. Trichoptera, ephemerids and other insects were collected.

After the dam at Dramhosso is finished, the water in the reservoir will probably be more acid than in the Brokopondo reservoir, because more humic acids are present in the Saramacca. At Afobaka the pH dropped after stagnation from 6.2 to 5.5. The initial pH at Dramhosso is 5.1. The composition of the plankton and oxygen conditions will be probably similar to those in the Brokopondo reservoir. There is no reason to believe that the waterhyacinth will develop less rapidly.

2e. TIBITI, COPPENAME AND TAPANAHONY (Tables 4 and III)

In order to collect data for comparison with characteristics of the Suriname River several other rivers in Surinam were visited occasionally. Samples were taken only once and therefore only preliminary conclusions can be drawn.

On January 25 and 26 the Tibiti River was sampled. This river is a tributary of the Coppename and it originates in the savanna. It is a typical Black Water. The stretch which was sampled is without rapids and forms part of the freshwater tidal zone of the river. The depth was about 6 m, the bottom was sandy and covered with decaying organic material. On the surface dense islands of waterhyacinth (*Eichhornia crassipes*) were floating, moving upstream and downstream with the tide. In the lower course, below Sabana, the black colour of the water turned gradually into redbrown and several kilometers before the confluence with the Coppename the water again changed into a greyish turbid colour, as is found everywhere in the coastal regions of Surinam. The vegetation along the banks changed also and mangroves predominated, indicating the zone of brackish water.

About one hour upstream from Sabana the river connected with a large swamp covered with a dense vegetation of waterhyacinth, which could be penetrated only with great effort. The roots of the waterhyacinths were crowded with many organisms, such as young fish, freshwater shrimps, crabs, larvae of dragonflies and waterbugs as *Ranatra*. Other waterplants like *Elodea*, *Utricularia*, and floating *Salvinia* and *Pistia* were present. Near the shore concentrations of *Montrichardia* were found.

The acid character of the water is shown in Table 4.
The pH is low and gradually increases downstream. The mineral

content also increases downstream and the same is found for the Cl-ion. When compared to the Suriname River in the tidal zone the water of the Tibiti appears to be much more acid. Evidently *Eichhornia* is not harmed by a low pH. This is confirmed by its presence in the even more acid Coropina.

The plankton of the Tibiti was not rich. It contained many iron bacteria and spiculae of sponges. Zooplankton was dominant. Rhizopoda and Heliozoa occurred together with the fungus *Actinospora*, and may be considered to be elements of the plankton from shaded streams. *Actinocyclus normanni* is a typical diatom from the freshwater tidal zone.

Large concentrations of the floating plants are found at locations with quiet water such as swamps. The *Eichhornia* in the tidal zone of the Tibiti must originate from the swamps upstream; the plants concentrate in the lower course of the rivers where they are carried passively by the current.

In the Suriname River there are no large swamps in the upper course and this may explain the comparative rarity of *Eichhornia* in its tidal zone. Waterhyacinths are adversly attacked by water current and also by wind acting on open spaces in stagnant water. pH and mineral content may be important factors for the growth of the plants, but there is no reason to believe that the waterplant will not multiply in the Brokopondo reservoir, if we compare the conditions in the Tibiti in this respect. Development in the Brokopondo reservoir confirms that stagnation of the water favours the development of *Eichhornia*.

The plankton sample from the C o p p e n a m e contained mainly sand and silt with many *Coscinodiscus*. *Coscinodiscus* is normally found in the brackish parts of the rivers. A list of the plankton species is given in Table III.

Attention is drawn to the fact that no snails were found among the waterplants, only a few naked slugs, living on the leaves of the waterhyacinth. Some shrimps were collected at Sabana, a.o. *Palaemonetes carteri* and *Macrobrachium jelskii*.

Various fishes were seen, mainly Characidae like *Gasteropelecus*. The river is known for its trapun fishing.

The T a p a n a h o n y River near Paloemeu was visited on March 8, 1964. The village of Paloemeu is located at the confluence of the Upper Tapanahony and the Paloemeu River. Samples were taken in the Upper Tapanahony and in the Tapanahony at the Maboegoe falls. Near French Guiana the Tapanahony joins the Maroni.

Like that of the Suriname River, the water of the Tapanahony was slightly coloured. In the Maboegoe falls Podostemaceae were abundant and a few plants of *Eichhornia crassipes* were found in quiet spots. Several *Pomacea* and *Doryssa* specimens were found on the rocks and also freshwater crabs.

The pH and electric conductivity were similar to those found in the upper reaches of the Suriname River.

The composition of the plankton community is very similar to that of the upper reaches of the Suriname River. The data should be completed, however, with observations at other locations and in different seasons.

The plankton at the Maboegoe falls contained many pieces of Podostemaceae, spiculae of sponges, *Alona, Cyclops, Chydorus, Rotaria, Lecane ludwigi, Arcella, Difflugia, Heliozoa, Centropyxis, Eudorina elegans, Dinobryon, Dictyosphaerium pulchellum, Kirchneriella lunaris, Microspora?, Mougeotia, Spirogyra, Closterium, Cosmarium, Gonatozygon mucosa, Micrasterias arcuatus, Pleurotaenium, Staurastrum trifidum, Oscillatoria, Diatoma, Eunotia asterionelloides, Nitzschia* (in moderate numbers), *Surirella*, veliger larvae, ephemerids, eggs of Odonata, chironomids and Simulidae.

3f. OTHER RIVERS (Table 5)

The road from Paramaribo to Albina crosses several rivers in the coastal region. On September 5 and 6, 1964, these rivers were sampled near the bridges or at the ferries. The sampling places were located in the tidal zone of each river. A few data on pH, electric conductivity and Cl-content are given in Table 5.

These data have some interesting aspects. The Suriname River near Paramaribo is brackish, but the Marowijne (= Maroni) at about the same distance from the sea is fresh. The reason for this may be that much more fresh water flows down the Maroni River than down the Suriname, which keeps the salt seawater from penetrating far into the estuary.

The Cottica and the Commewijne, both narrow meandering rivers, have a very low pH, though the salinity is relatively high. This aspect is very interesting. It may be caused by the paucity of lime in the subsoil and by a large amount of dissolved humic acids carried down by the rivers, both of which must be characterized as Black Waters. Research, recently carried out on the structure of the subsoil in the coastal regions (PONS 1966), has established the presence of tropical podsols with a low pH. It has been mentioned before that Black Waters originate from these podsols, which are often found in white sand savannas. It would be interesting to know more about this unique type of tidal water.

The plankton at the sampling stations in the rivers is composed either of species normally found in freshwater tidal zones, as is the case on the Maroni, or in the more brackish tidal zone, as is the case in the Cottica and Commewijne. The colour of the water was clear in the Maroni; especially at low tide much detritus was present. The Cottica carried blackish-brown detritus and was opalescent, probably due to a high content of iron. The Commewijne carried blackish-brown detritus and also flaky organic material. The water colour was greyish brown and the turbidity was high.

The following plankton organisms were found in the Marowijne: *Bosminopsis, Diaptomus, Eudorina elegans, Dictyosphaerium pulchellum* (moderate numbers), *Kirchneriella obesa, Cosmarium, Cosmocladium, Staurastrum, Actinocyclus normanni* (moderate numbers), *Melosira granulata* (moderate numbers), *Pinnularia, Surirella,* spiculae of sponges. At low tide on the next day many *Actinocyclus* were present, with calanoid copepods, veliger larvae, *Bosminopsis deitersi, Cyclops, Moina, Eudorina elegans, Dictyosphaerium, Closterium, Euastrum monocylum, Bacillaria paradoxa, Melosira granulata, Surirella* (moderate numbers), spicula (moderate numbers). This plankton is typical for the freshwater tidal zone. No salt or brackish water organisms were found.

In the Cottica also organisms from the freshwater tidal zone were present, but the diatom *Coscinodiscus* may be considered to be a representative of brackish water. *Bosminopsis deitersi, Cyclops, Cathypna luna, Lecane, Monostyla bulla, Arcella, Euglypha, Dictyosphaerium pulchellum, Mougeotia, Spirogyra, Closterium, Desmidium, Micrasterias arcuatus, Actinocyclus normanni* (moderate numbers), *Bacillaria paradoxa, Coscinodiscus, Diatoma, Eunotia,* many *Surirella, Synedra, Actinella* (moderate numbers), spicula of sponges.

In the Commewijne a brackish water plankton was found with many broken parts of *Coscinodiscus*; calanoid copepods, *Actinocyclus normanni, Actinoptychus, Coscinodiscus* (moderate numbers), *Surirella,* spicula.

The lower Suriname River was sampled regularly by VAN DER HEIDE (1965–1967).

2g. RIVULETS AND SMALL STREAMS (Tables 6 and IV–VI)

Most inlands parts of Suriname are slightly undulating, with watersheds not more than a few hundred metres above the level of the river.

Flat parts throughout the forest generally become sodden or flooded in wet weather. Such places are found especially in the lower parts of the forest near the river. The small forest streams are fed either from these flooded areas or by rivulets flowing from the higher grounds. These rivulets are numerous. On the higher grounds they are small with a swift current; near the river they are wider and sluggish. As long as the rivulets and small streams are small, the canopy above them remains closed often right down to the river. When the canopy is broken, the water of the stream is exposed to almost full sunlight.

The subsoil of the forest is composed of laterite and sand. It is covered by a thin layer of topsoil and decaying leaves. The bottom of the small streams is usually sandy and in the slower parts muddy.

As the rivulets feed the smaller streams and these in their turn empty into the river, it is interesting to have some information on the character of the water.

For this purpose a few rivulets, crossing the road to B r o w n s w e g were analyzed in November (Tables IV and V). The rivulets generally show a lower pH, a higher conductivity, a lower temperature and a lower oxygen content than the Suriname River itself. The water was clear and here and there exposed to sunlight. Just near the road, filamentous algae were found (*Mougeotia, Spirogyra, Zygnema, Batrachospernum*). Between these many desmids and diatoms were found and also many rotifers. Characteristic for this dark environment with decaying leaves is the fungus imperfectus *Actinospora*. In stagnant places in these rivulets fish and tadpoles were found.

The rivulets mentioned here feed the Compagnie Kreek and the Wedang Kreek, which enter the Suriname River north of Afobaka. The water of these small streams is also clear and the chemical analyses and the composition of the plankton did not differ much from those of the rivulets. The small streams run slower, the water is deeper and as a rule the conductivity is somewhat lower.

Two other small streams are found in the region of the future lake, South of K a b e l. The water was also clear, but on the day of inspection there was no current. They were composed of isolated stagnant pools and probably only contain running water after rains. The temperature of the water was as low as in the forest rivulets. The Witte Kreek, however, showed a surprisingly high electric conductivity. The samples were taken not far from the mouth of the stream where there may be some influence of river water. This is confirmed by examination of the plankton. Many species from the river were found: iron bacteria (*Leptothrix*?), Cercariae and *Actinospora* and in some parts of the other stream *Spongilla, Plumatella* and *Rhipidodendron huxleyi*.

The K a s s i e K r e e k near Pokigron was also nearly dry. In a pool filled with clear water a low electric conductivity was found. This stream drains into the

Suriname River during wet weather and is located just south of the future bank of the lake. An isopod collected here looked similar to *Livoneca symmetrica*.

Finally, the M a c a m b i K r e e k, located east of Brownsweg, runs southward into the lake basin. The conductivity and the pH are higher than in the other rivulets and streams and Characeae were found. The water is clear and it is used as a water supply for the military camp of Brownsweg.

Summarizing, we may say that in the lake region all rivulets and small streams contained clear water. The mineral content was variable on account of variations in the subsoil. Black Waters are not found in this area.

The Black Watertype is nevertheless found in Suriname. The Suriname River itself has a tributary of this type. Such waters are found downstream in the savanna region. The Carolina Kreek near Zanderij, the Coropina Kreek near Republiek, and the Cola Kreek near Zanderij are examples. The pH in these waters was found to be very low (pH 3.3–4.0) generally, due to dissolved humic substances which give the dark brown or black colour to the water. The plankton did not differ much from other streams mentioned and a fair number of aquatic insects (*Ranatra, Belostoma*), fishes and several young crabs (*Trichodactylus*) were found. The presence of so many fish and even crabs and freshwater shrimps in this very acid and soft water is rather surprising. In acid waters in temperate climates fish and larger crustaceans are absent. It would be interesting to know more about the ecology of the organisms present.

In one of the larger Black Water streams, the C o r o p i n a K r e e k, a low pH was found (Table 6), and the conductivity was relatively high.

Organisms found were: *Alonella dadayi, Pleuroxus, Scapholeberis mucronata, Cyclops, Arcella, Rhipidodrendron huxleyi, Dinobryon, Lecane, Desmidium elegans, Cosmarium, Hyalotheca indica* (abundant), *Spirogyra, Diatoma, Synedra?*, and chironomids. On the floating leaves of *Nymphaea*, only the snail *Acroloxus lacustris* was found. In several places beds of *Marsilia quadrifolia, Salvinia, Elodea* and *Eichhornia* had developed.

At the Coropina Kreek station a large flooded area was found; depth about 0.5 m. It had an open connection with the stream and was probably only flooded during the wet season, as the submerged vegetation of grasses (*Lolium*) suggested. The water showed a pH of 5.7 and a conductivity of 28 μS. The μS and pH values are different from those found in the Coropina (Tables 6 and 7). The water was also very clear. The opinion seems justified that Black Waters such as the Coropina receive their water mainly from podsols in the savanna. Very few organisms are found in the swamp. The plankton contained: *Alonella, Diaphanosoma, Arcella, Dinobryon* (numerous), *Mougeotia, Euastrum, Hyalotheca, Staurastrum*, and *Pristina longiseta*.

During two visits to the upper reaches of the Suriname River a few rivulets and small streams, south of the future lake region, were sampled.

The Sipari Kreek near Aurora was flowing slowly, in contrast to the Anjanwoyo Kreek near Botopassie, which had a fast current and turbid brown water. The Anjanwoyo Kreek had a higher conductivity than the Suriname river, but the Sipari Kreek showed hardly any difference (Table 6). Near Ligolio two rivulets with clear water had nearly the same conductivity and pH as the Gran Rio. The temperature was somewhat lower, the water being shaded by trees. The current was fairly slow.

In a rivulet near Assidonhoppo a higher conductivity was found than in the Pikien Rio. The composition of the plankton did not differ much from that found in the other rivers in this region.

The few observations show, that there is some variation in conductivity in the various streams and rivulets. Generally the conductivity of rivulets and streams is somewhat higher than that of rivers.

3.   STAGNANT WATERS

(Tables 7 and VII–VIII)

Stagnant waters such as swamps are rare in the interior of Suriname, they are difficult to reach because of their isolated situation.

In the region of Afobaka there were several ponds, recently constructed during the building of dams and roads. These ponds contained no aquatic macrophytes, the bottom was muddy and many dead trees were standing in the water. The trees had died as a result of permanent immersion in water that was no more than 1 or 2 m deep. Many trees along the river are completely covered by water during floods for weeks on end and do not seem to be damaged. Anaerobic conditions caused by reduction processes in the mud might explain the eventual death of the trees during permanent immersion.

In a few ponds some vegetation of grasses was found, especially when they were situated in open spaces. Probably they were originally natural swamps. In the forest pools or swamps no vegetation was developed because of lack of light.

The water in the ponds and swamps was turbid and its colour was yellow, grey or green. Most of the pools contained clear water, but during heavy rains the pools along the roads might be coloured red by washed-in laterite soil. In the tables the colour is indicated. The grey or yellow colour is caused by suspended material of unknown origin. A green colour implied the presence of many algae.

The temperature of the water was higher than in the river. In sun-exposed swamps temperatures up to 34°C were recorded. This high temperature was also found at the surface of the part of the river that was dammed. Fluctuations of the daily temperature were not recorded, but we might expect similar fluctuations as were found in the river after damming.

The oxygen content was nearly always low. Large fluctuations during the day may occur, but this phenomenon has not been studied further. In a pool without vegetation at the base of a dam, the water was nearly saturated and in two other cases fairly high values were found. This, however, is not the normal picture. In these shallow waters containing large quantities of organic material a low oxygen content might be expected.

A list of the plankton found, is given in Tables VII and VIII.

Especially in the turbid swamps and ponds unicellar flagellates were abundant, e.g. *Euglena* and *Trachelomonas*. In the sun-exposed grass swamps the flagellates were less numerous and green algae, such as *Pediastrum duplex, Coelastrum cambricum* and Desmids such as *Micrasterias brasiliensis* predominated. Brighter light conditions determine the character of the biocommunity here.

In these shallow ponds, swamps and pools the conductivity was higher than in the river and the small streams. In some pools located just at the base of dams in which no vegetation was present, a red clay bottom was visible and very high values were found (100–210 μS, see Table 7). When, however, this red clay was agitated in a jar together with rainwater, such a high conductivity was not found. Therefore, the enrichment of the water of these pools with minerals must have another origin. As rainwater has a very low conductivity (7 μS) the source may be seepage or a well. In Table 6 such a well is recorded. Here iron is the cause of a high conductivity.

In comparison the conductivity of the Suriname River is low, but after impoundment of the river the conductivity increased to values comparable to those of the shallow ponds and swamps in September 1964, especially near the bottom of the lake.

There is a remarkable difference between the pH of the ponds and swamps and the impounded lake. In the latter the pH dropped from about 6.5 to 5.5 after the stagnation. In the ponds and swamps the pH remained about the same as in the running water of the Suriname River. In Guyana, CARTER (1934) found pH as low as 4.4–4.5 in grass swamps and forest pools that never dry up. In this case the water was coloured black or brown and the high acidity was caused by humic acids. The pools and swamps at Afobaka mentioned above, are not permanent and the water is not coloured by humic acids. As a result the pH is higher. In this connection it is understandable that the clear pools at the base of the dam which have a high conductivity also have a relatively high pH as no humic acids or decaying organic matter is present. In another chapter we shall see that the water of the impounded lake is coloured brownish by decaying material of the drowned forest. The humic substances in this case may cause a low pH of the lake.

In all probability most of the ponds and swamps mentioned had developed recently. This short period of time may be the reason why no humic substances are present yet. In Table 7 a few older swamps are included; their pH was lower (5.1, 5.7).

The swamp at Pokigron is an old branch of the Suriname River. This stagnant water has a low conductivity and many waterplants such as *Eichhornia crassipes*, *Limnanthemum*, *Lemna valdiviana* and *Utricularia* are present. The presence of vegetation indicates that this body of water is more permanent, and older than the swamps at Afobaka. The amount of decaying organic matter derived from the vegetation is larger. More humic substances and CO_2 are present and the pH will be correspondingly lower.

Fish, shrimps and crabs could be found in all the swamps mentioned. Most of the smaller fish species were surface dwellers. The larger species lived at the bottom and came to the surface of the shallow water in order to replenish their supply of oxygen by intestinal respiration.

4a. Daily observations before and after the closing of the
    dam (Figs. 7–9; Tables 8–9)

In tropical countries climatological variations take place during
the day as they do in temperate parts of the world. Therefore temperature, oxygen, pH and conductivity in the river were measured
in the course of a day. The first series of observations was made on
November 18, 1963 just near the bank. But work at the dam influenced the results, especially the values for conductivity. So the
sampling station was moved to the middle of the river on December
6. Samples were taken at the surface and at a depth of 4 m. Considerable daily fluctuations were found at the surface. But at a depth of
4 m these fluctuations were less pronounced. These observations
were followed by a vertical series of samples in the morning and in
the afternoon at a depth of 0, 1.5, 2.5, 4.5 and 6.5 m. In May, June
and July when the water had been impounded for a considerable
period these observations were repeated not twice but three times
a day.

Before the closing of the dam (Figs. 7–8)

Daily observations took place from January 14–23, during the
first days the waterlevel dropped. In the afternoon the temperature

32

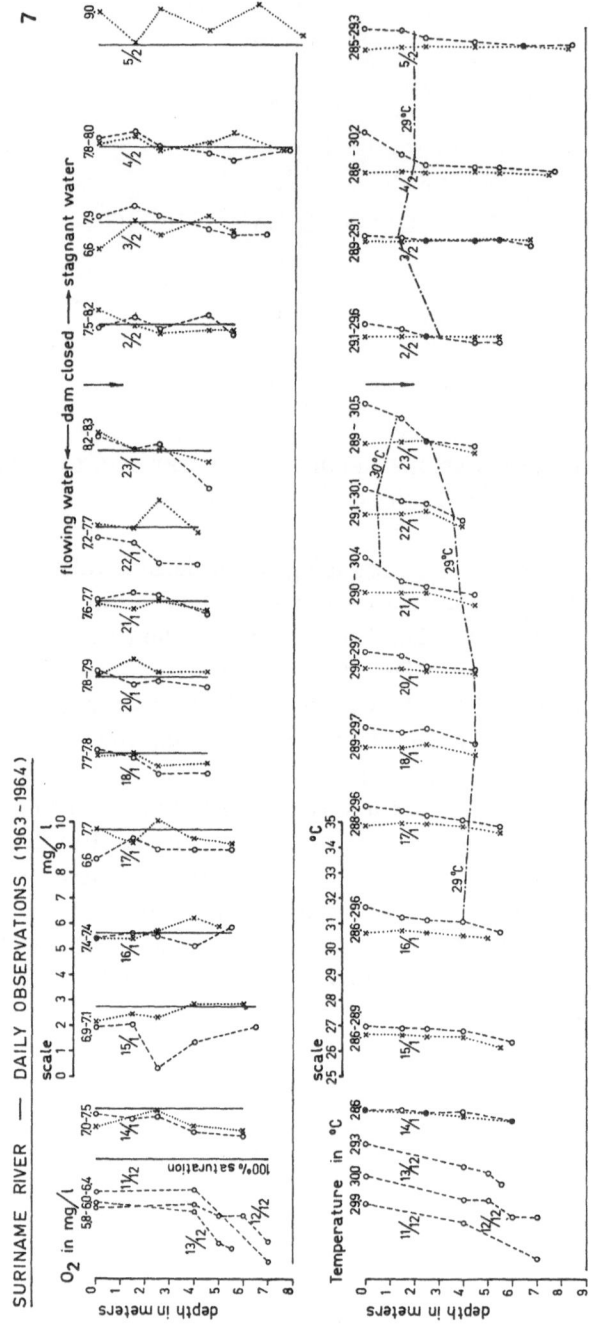

SURINAME RIVER — DAILY OBSERVATIONS (1963-1964)

7

33

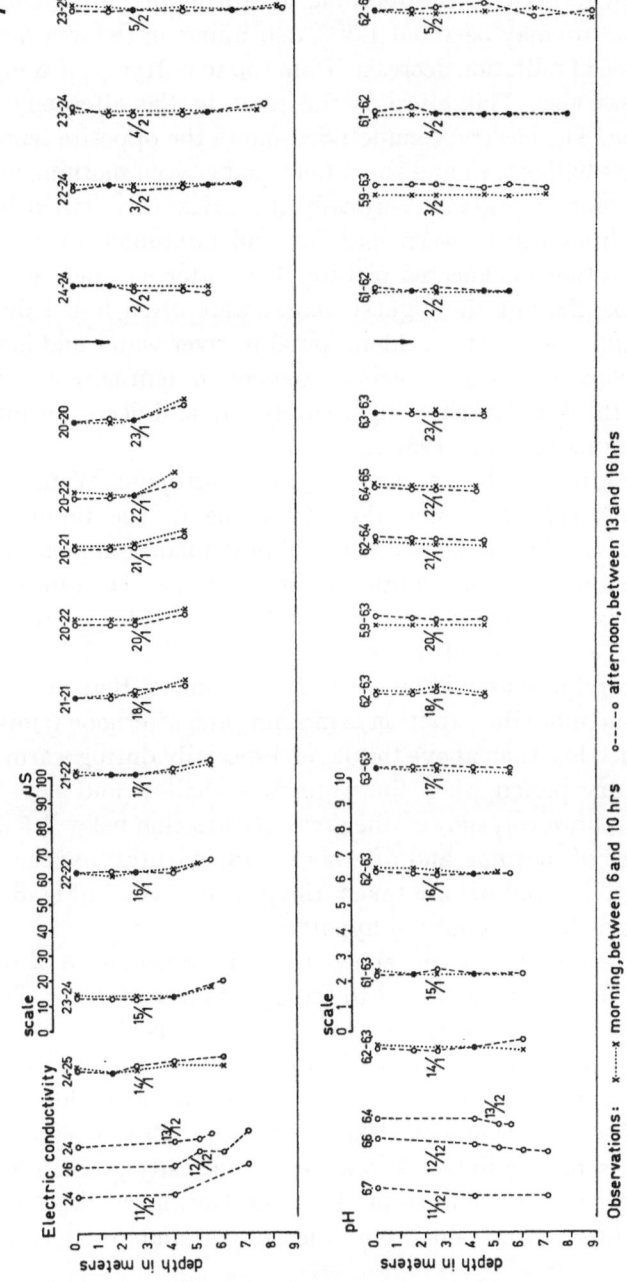

7

Observations: x------x morning, between 6 and 10 hrs   o----o afternoon, between 13 and 16 hrs

was higher than in the morning at all depths. Difference in the temperature may be about 1.5°C, depending on the weather. As the influence of radiation decreases from top to bottom, the temperature decreases also. This effect is the same in the afternoon and the morning. The electric conductivity shows the opposite tendency. It increases with depth and the difference between morning and afternoon is not very great. Vertically the pH hardly varied but there was a difference between morning and afternoon. Oxygen finally gave a rather unexpected picture. The water was nearly saturated at all depths but the highest values were often found during the morning. This effect is seldom found in river water and is not easy to explain. In eutrophic rivers the oxygen content normally increases during the day by increasing assimilation, and decreases during the night (SCHMASSMANN 1951).

The Suriname River, however, is oligotrophic. Waterplants are absent, except for some Podostemaceae in the rapids. Oxygen productivity by the small amount of phytoplankton present is negligible and oxygen consumption is also very low. The sandy or rocky bottom is clean. Apparently, abiotic factors to a large extent control the oxygen content of the water.

On careful examination of Fig. 7 it appears that at a depth of 2.5 m and more the variation in morning and afternoon temperatures was much less than above this layer, especially during warm weather in the dry period when the water was shallow and slow-running. Oxygen, however, showed the greatest variation below 1.5 m. If the averages of morning and afternoon temperatures, oxygen content, conductivity and pH are taken, the picture given in Fig. 8 appears. This leads to the following hypothesis.

The temperature curve shows that the warming up of the water during daytime is restricted to the upper layer of 2.5 m. This is the effect of direct absorption of radiation which is always restricted to this top layer. Under tropical circumstances the rays of the sun reach the water at an almost vertical angle in the middle of the day and reflection is minimal. When the temperature rises the density decreases progressively. A few degrees already constitute a great difference in density in tropical waters. During the day the warmer and lighter water remains at the surface and resists the mixing efforts of weak horizontal currents. The warm layer of water on top prevents exchange with the water below the 1.5 m mark. During the night the surface water cools and sinks down. These vertical movements often fail to mix the water completely with the possible result that early in the morning more oxygen is found in the deeper and colder layers. Further examination of Fig. 7 shows that pH and conductivity are also controlled by the same mechanism which has its critical depth at 1.5 m.

Absorbed radiation does not result in the formation of a proper stable thermocline. In the morning the temperature becomes uniform and there is no thermal stratification at all and in the afternoon the temperature increases above a certain depth. This effect was even more pronounced after the dam was closed and the water had been stagnant during a prolonged period.

Stable thermoclines are formed in some tropical rivers. BRAUN (1952) has found a thermocline in the deep slow moving water of the Rio Tapajoz. In the Suriname River the author found a stable thermocline during a period of several days under high water conditions (see Fig. 7).

AFTER THE CLOSING OF THE DAM (Fig. 9; Tables 8–9)

The observations in the morning and in the afternoon were repeated after the dam was closed, on February 2, 3, 4 and 5. The river became stagnant and the waterlevel rose swiftly. Oversaturation with oxygen was then often recorded from top to bottom. In the afternoon the temperature decreased from the surface downwards but in the morning temperature and also conductivity and pH were always uniform.

The most important phenomenon during this period is the oversaturation with oxygen. However, it does not reach the high level that is found in eutrophic water with plenty of algae such as Lake Volta (EWER 1966). The observation on February 5 shows maxima at various depths resulting from the diurnal cycles. Here the same mechanism may be at work as described previously with the only difference that as soon as the water became stagnant, the few reductive substances sedimented. As a result oxygen was not consumed anymore in the deeper layers and oversaturation resulted (see also SCHMASSMANN 1955, MORTIMER 1956, and LINDROTH 1957).

In May, June and July when the water in the lake had been stagnant for a long time and a rich plankton community had developed, the drop in temperature and oxygen at a depth of 1.5 and 2.5 m became even more pronounced (see Fig. 9). Thermoclines exist only temporarily. Oxygen is absent below 2.5–3.5 m. As in the river, the effect of radiation is restricted to the upper layers. Penetration of light is also limited by the dark colour of the water and as a result the temperature drops sharply.

Wind and horizontal currents have no mixing effect. At Kabel where the effect of the wind is greater, the decline in temperature and oxygen is less pronounced and reaches deeper layers.

BRAUN (1952) found thermoclines in shallow lakes in the Amazon region also at 2–3 m. He did not check the diurnal cycle but instability is suggested by the fact that he did not find thermoclines on several occasions. RUTTNER (1962) found thermoclines at a considerable depth in deep lakes in Indonesia and emphasizes the stability. In some lakes in tropical Africa no thermoclines are developed at all (DAMAS 1939, WORTHINGTON 1930).

CARTER (1934) observed temperature inversions in the shallow swamps of Guyana and vertical mixing. The present author also observed inversions and mixing at least of the superficial layers.

The upper 2 or 3 m of the water are strongly influenced by weather conditions. However, the cooling and warming up of this water does not result in the formation of a stable thermocline. The vertical mixing forces are not sufficiently strong. Sometimes the weather conditions may cause a drop in the temperature late in the afternoon. This happened on May 23 and 25, when heavy rains started in the afternoon and lasted until the evening. Oxygen content dropped in a parallel manner to the temperature.

Generally speaking the amplitude of temperatures in May, June and July is greater than in January and February. But below 4.5 m the variations are equal or even smaller. The same applies to oxygen conditions. In the upper layers of the water the greatest diurnal variations in temperature and oxygen content are found.

8

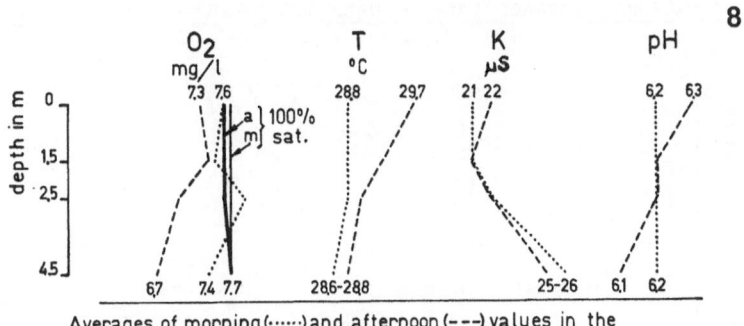

Averages of morning(·····)and afternoon(– – –)values in the
Suriname river at Afobaka (14-23.I.1964)

The plankton community changes to resemble the community
found in shallow swamps. The higher temperatures may favour the
development of green organisms that produce oxygen. The BOD
values also rise and during the night oxygen content will soon be
exhausted. Oxygen content is always higher in the afternoon in
contrast to the situation in the river.

On July 16 and 21 the pH was somewhat lower in the morning
than in the afternoon and this observation was made quite often on
other occasions. This happened throughout the watercolumn while
oxygen and temperature underwent changes only near the surface.
It should be caused therefore by chemical processes going on at all
depths. The changes in conductivity should sustain this theory. The
observations on July 20 illustrate the changes in electric conductivity. On this day and on several other occasions (see Figs. 7 and 9) a
significant increase in conductivity was found at a depth of 2.5 m
at 10.00 h. In the afternoon at 14.30 this increase was found at
4.5 m and late in the afternoon it had disappeared. Near the surface
during the night decomposition of organic material will increase the
mineral content, but during the day the minerals might have been
taken up by growing plants and algae. Decomposing organic material
will sink down in the course of the day enriching with decomposition
products the layers it passes through.

In tropical waters the decomposition of organic material is rapid
and in deeper water the organic matter is mineralized before reaching the bottom. Depending on the weather a pulse of dissolved

38

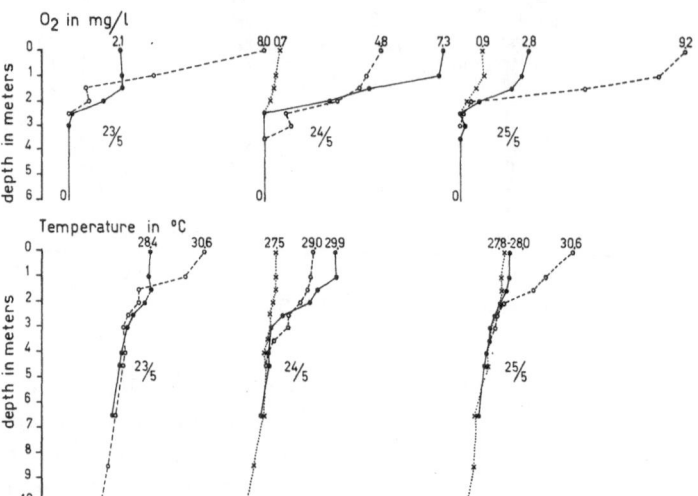

Observations

x·········x   morning   between 6 and 10 hrs

o- - -o   afternoon between 13 and 16 hrs

•——•   afternoon between 16 and 18 hrs

horizontal scale O$_2$ in mg/l

0 1 2 3 4 5 6 7 8 9 10

horizontal scale temperature in °C

25 26 27 28 29 30 31 32 33 34 35

horizontal scale electr. cond. μS

0 10 20 30 40 50 60 70 80 90 100

horizontal scale pH

0 1 2 3 4 5 6 7 8 9 10

9

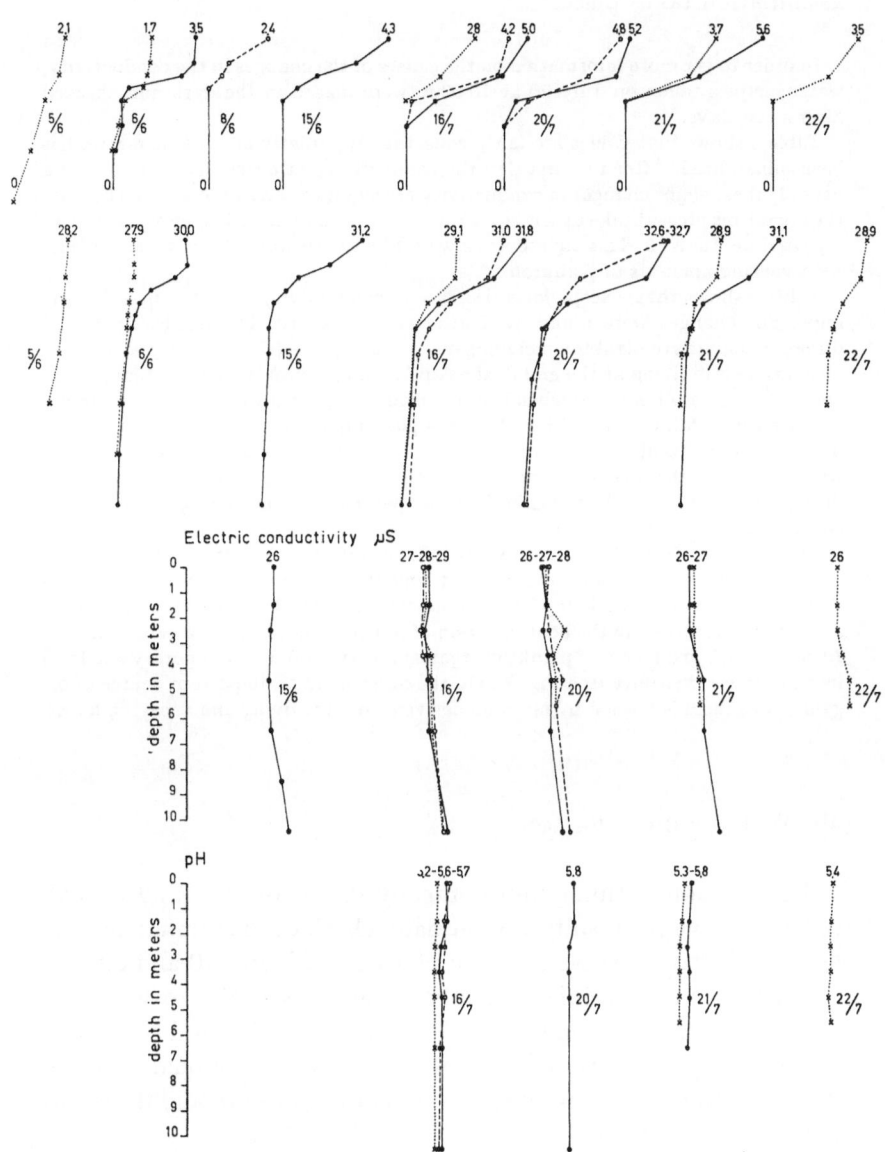

minerals travels downwards apparently during the hours at which
assimilation takes place.

In order to get more information on the causes of the changes in the conductivity,
water samples taken on July 20 at 10.00 h. were placed in the dark and checked
after a few days.

Table 8 shows that values are fairly constant. Apparently all organic matter has
been mineralized. After the first day the conductivity at every depth decreased a
little. If these slight changes in conductivity are significant at all, it may be because
the organisms present take up a certain amount of the minerals. This experiment was
repeated on June 16. This time bottles were filled with water from various places
with varying amounts of plankton.

Table 9 shows that, except for a decrease in conductivity after the first day, no
important changes were found even after several weeks. The bottles contained
different amounts of plankton including many crustaceans (*Cyclops*, *Diaphanosoma*),
which were still living at the end of the experiment. Concentrations of filamentous
algae (*Spirogyra*, *Mougeotia*) when kept in an aquarium died within a day. Therefore
a bottle with a fair amount of the algae was placed in the dark, and a rapid increase
of conductivity resulted. Decomposition was very fast and a strong smell of H_2S
was noticed. In the lake a great many filamentous algae developed in quiet places
during the dry season; they died in the rainy period and thus increased the mineral
content of the water.

This decomposition may occur locally and the effect may vary greatly from one
day to another. Large concentrations of filamentous algae were found at a certain
depth one day near Koffiekamp for example. Because of lack of light these algae died
and decayed, increasing the concentration of minerals, but only at a certain depth.
Also the death of a bloom of plankton organisms may produce a temporary and local
increase in conductivity (see Fig. 9). On the other hand an important source of organic substances is formed by material derived from the dying and decaying forest.

4b. WEEKLY OBSERVATIONS

The Suriname Aluminium Company (Suralco) was kind enough
to let me have the results of their daily check of the rainfall and the
level of the river. These data are plotted in Figures 10 and 30.

The total annual rainfall at Afobaka varies from 2000 to 3000 mm;
May and June are normally wet, September, October and November
are dry. In 1964, however, the weather was exceptionally dry in
April and May. Extensive bushfires swept the swamps and the coastal woods of Surinam.

The river levels are expressed in NSP (New Surinam Level =
average sea level). The rise of the water level after the closing of the

dam and after a period of rainfall in December in the river can be seen in Fig. 10.

Station 1. *Afobaka* (Figs. 10–11; Table IX)

In Fig. 10 the weekly observations of temperature, conductivity, pH and transparency of the surface are given.

Plankton was sampled by means of a plankton net drawn through the water (Table IX).

From November 1963 until February 1, 1964 it was still possible to collect data in the original Suriname River at Afobaka. The sampling station was situated in the middle of the river about 300 m upstream of the dam. In the dry season the river was about 150 m wide, 4 to 5 m deep and the current was slow.

In the wet season the river is much wider, the banks are covered with water and large areas bordering the forest are flooded. The water level rises more than 5 m. As a result low trees disappear completely during a certain period. The trees stand this temporary drowning without harm. Trees growing on a higher level were less adapted to floods than trees bordering the rivers and streams. After the closing of the dam the adapted trees for a long time formed a green belt though the inner forest was already leafless and dead.

Immediately after the dam was closed the water rose 5 m in the first week, 3 m in the second and 1 m in the third. The river flooded the bordering forest and this caused the rate of rise to level off. In the week before March 25 heavy rainfall caused a swift rise of the water level and at the end of May with the start of the main rainy season the level kept rising fast for a prolonged period.

Before the dam was closed the river water was nearly saturated with oxygen. At the end of the first week after the closing of the dam oversaturation was recorded, the amount of plankton was small and the degree of oversaturation was not great. Oxygen production by phytoplankton could not be detected in tests. Sedimentation of reductive substances during this period lowered the BOD and as a result more oxygen remained. The actual cause of this oversaturation must be abiogenic.

CHEMICAL ANALYSES AND WATER LEVEL OF THE SURINAME RIVER AND THE BROKOPONDO LAKE (1963-1964)

10

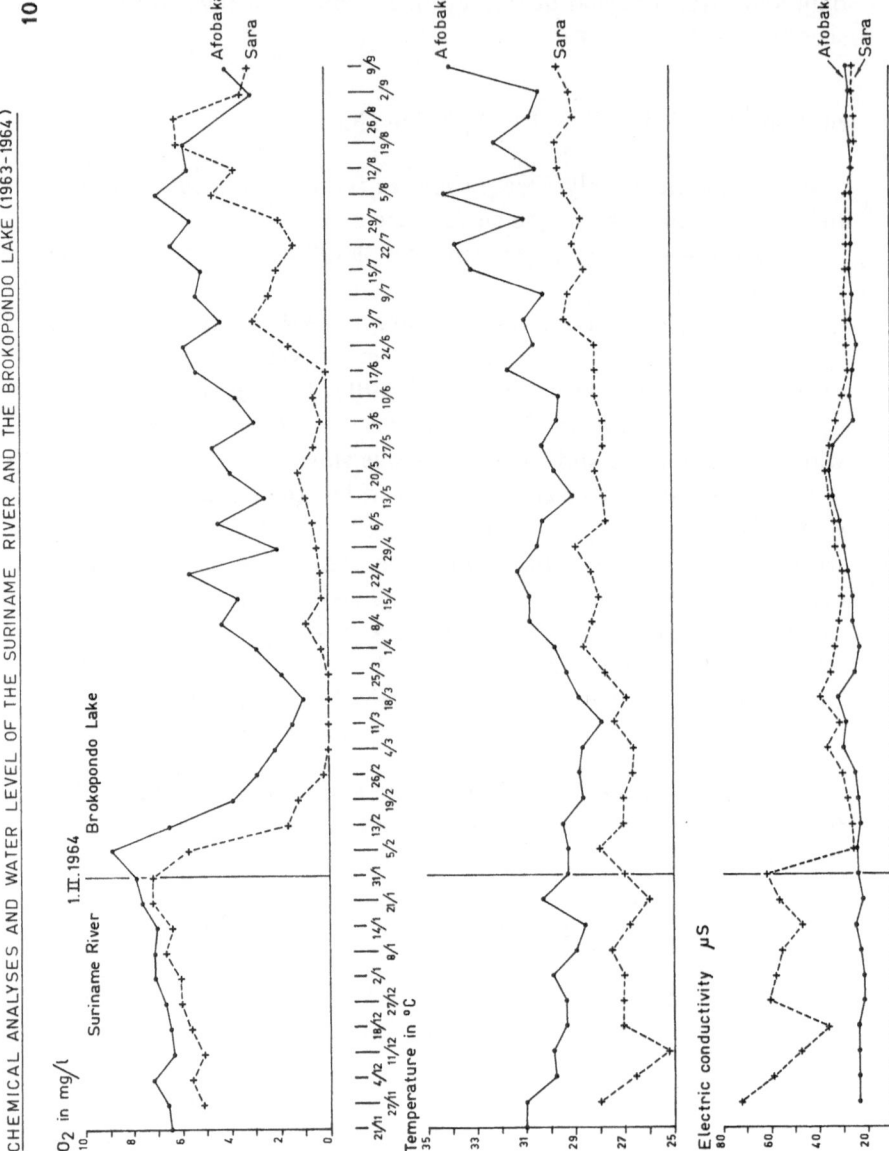

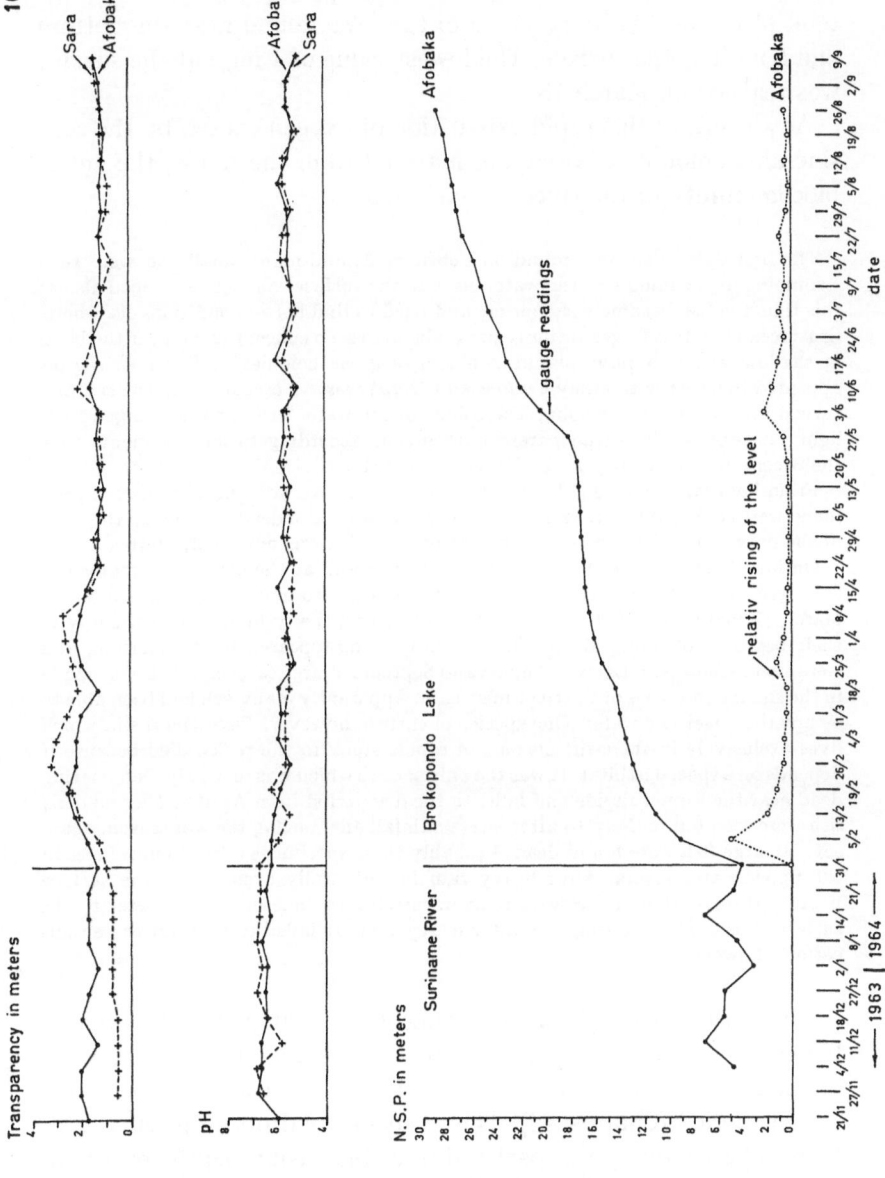

10

After the first week of stagnation a steep drop in oxygen content took place: on February 17 no oxygen was found near the bottom and 5 mg/l at the surface. The lowest value of 1 mg/l at the surface was reached on March 18.

As a result of this rapid exhaustion of oxygen caused by the considerable amounts of decaying material from the forest, the entire biocommunity of the river was affected.

The first dying fish were found on February 28 and many small fish were seen swimming in the middle of the water close to the surface. As H_2S developed also at this time the fish became very uneasy and tried to find better conditions elsewhere. Crustaceans such as larger shrimps and crabs were seen concentrating near the bank in shallow water. A phytoplankton bloom of green colonies of *Eudorina elegans* appeared together with many *Cyclops* and *Diaphanosoma brachyurum*. The samples turned into a thick green soup. The colour of the water in the river changed from light turbidbrown into transparent dark green. According to measurements with the Secchi-disc the transparency increased slightly.

From February 28 dead fish could be found nearly everyday near the dam. Specimens were collected by dr. M. BOESEMAN. There were also dead fish among the trees of the drowned forest. The total amount of dead fish was not great, however.

On March 11 the bloom of *Eudorina* vanished and at the same time many fish appeared at the surface, gasping for air. There was also more rain than before. On March 12 many more fish came up, mainly catfish, many hundreds of them with their beards protruding above the water. They disappeared in the morning and were never seen again. Later in August and September large catches of fish were made in the small tributaries of the river upstream. Apparently many fish had been able to escape the anaerobic water. One species of catfish, however, *Plecostomus* sp., which lives exclusively in the swift current of rapids stuck to the rocks, died because it depends on a special habitat. It was the only species which was regularly seen floating dead near the former rapids and falls. In the dry period from April to May no dead fish were seen but on May 14 after heavy rainfall announcing the wet season, again several large fish were found dead. Probably these specimens came from refuges in the upper water layers. After heavy rain fish generally come up to the surface because the lowering of the watertemperature brings anaerobic deep water to the surface. Observations during and after heavy showers failed to bear out these indications, however.

The period of rapid oxygen exhaustion, changing transparency, bloom of *Eudorina elegans*, decrease of temperature, increase of organic matter and minerals is defined as the transition or metatrophic period (Chapter 4d). The temperature dropped gradually to 28°C. The decrease was partly due to increasing depth, partly to cooler water coming from the drowning forest mixing in. The water under the canopy of leaves and branches was about 2°C colder than that of the river. Because of the large amounts of decaying material from the forest, the conductivity increased. The pH dropped to 5.5, and the presence of H_2S was established by smell.

During the transition period the plankton community changed. Certain components of the original river plankton increased, as *Eudorina elegans* and zooplankton, others disappeared, like *Eunotia asterionelloides*. The latter may have disappeared through sedimentation when the water became stagnant. Several other species with a relatively high specific weight also sank to the bottom. In this transition period those elements of the river plankton which normally react favourably to disturbances are most likely to increase in numbers, e.g. *Eudorina elegans*. In the river it appears at the beginning and at the end of the rainy period when the water is disturbed and there is a temporary increase in organic matter. This temporary increase of *Eudorina* has already been mentioned at Pokigron. It always appears in the metatrophic zone. Near Afobaka it multiplied in the transition period, but when the stagnation became permanent the process of biological selfpurification failed and *Eudorina* disappeared. The metatrophic zone moved from Afobaka to the south and the green "wave" of *Eudorina* moved in the same direction. This zone could easily be followed thanks to the typical green colour.

After the transition period the oxygen content in the surface layers was re-established. A new plankton community was formed composed of flagellates, rotifers and cladocerans. From March 18–25 heavy rains raised the waterlevel.

This rise of water level was accompanied by a rise in temperature, a drop in conductivity and fluctuations in pH at all levels. The transparency increased temporarily and river plankton with species such as *Eunotia asterionelloides* appeared from the south. The horizontal transport of river plankton proves that the water of the lake is influenced mainly by the river. Influence of rainwater must be less important as at this time the total surface of the forming lake was not very great. Some effect of dilution by rainwater may be concluded from the low temperature on March 11. However, it affects the surface and the 1.5 m layer only while it causes some decrease

46

PLANKTON OF THE BROKOPONDO LAKE AT AFOBAKA (1963-1964)

11

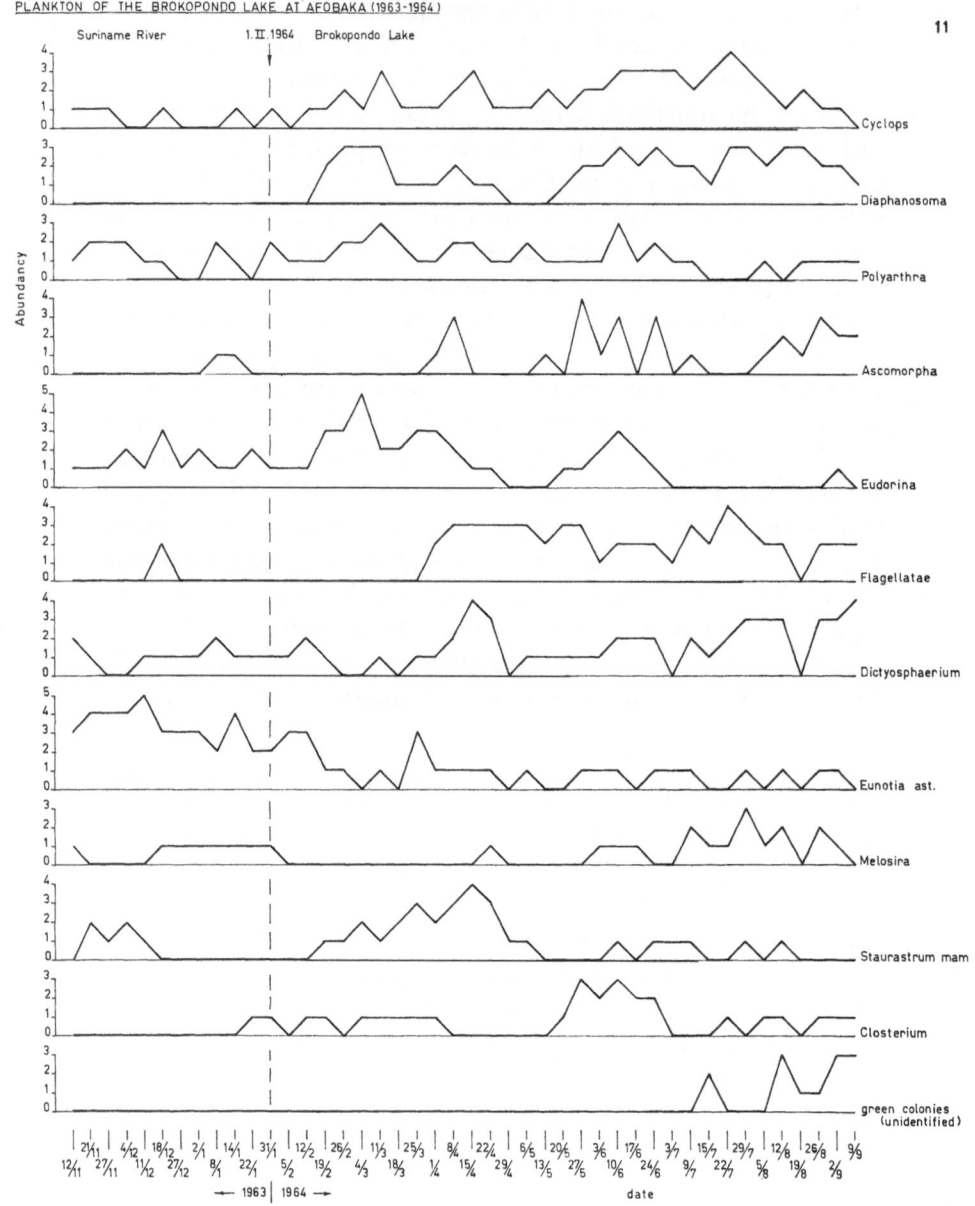

in conductivity also, but the causes of the events are difficult to analyze without detailed observations of watercurrents.

The effect of the short rainy period did not last long, as was shown by the rapid decrease of *Eunotia* and *Eudorina*.

In the period from March 25 until April 22, an increase of flagellates, *Dictyosphaerium pulchellum*, desmids and *Cyclops* was accompanied by a rise in oxygen content, temperature and conductivity. The plankton was present only in the top layers down to 3 or 4 m where oxygen was present. Colonies of the rotifer *Sinantherina spinosa* were characteristic of this plankton, showing up everywhere as white spheres. This species develops well at low oxygen values. Chlorophyceae and desmids benefitted from the fact that more light penetrated the surface of the water as many trees drowned and the open water area became enlarged.

From the beginning of April until the end of May the weather was dry. The water was no longer disturbed or influenced by rain as was demonstrated by the constantly low transparency and the gradual increase in conductivity. The temperature, however, was decreasing. This was unexpected and is still unexplained, as there was more sunshine. Daily fluctuations reached a maximum.

On the whole the oxygen content was lower than in the former river. The surface layers showed great variations in oxygen contents from week to week, from day to day, and even from hour to hour; the oxygen closely followed the fluctuations in temperature. In the plankton community a sudden drop in numbers of crustaceans, desmids and *Dictyosphaerium* was noticed but the flagellates, especially *Trachelomonas* and *Strombomonas ensifera* remained abundant.

At the end of this dry period a brown, flocky detritus appeared in the surface layers containing many iron bacteria. The biocommunity was very much like the one found in saprobic water. In the forming lake the amount of decaying organic matter originating from the dying forest had increased to such a degree that exhaustion of the oxygen content was permanent except for the surface layers where numerous unicellular flagellates produced oxygen. Swamps with dead trees near Afobaka showed a comparable biocommunity. The plankton in the lake developed mainly in 3–4 m of surface water and may be described as swamp plankton. The greater part of the lake will become shallow and swampy; most dead trees will remain for years. The swamp plankton, found in the open water of the lake, will eventually be completed with tychoplankters and invertebrates.

The first disturbance in the dry period occurred in the night of May 13–14. Dead fish were found near the dam in the morning.

The week before May 27 the water level began to rise more rapidly, marking the beginning of the long wet season. After a sudden period of rain in the week of June 3 the water level rose less quickly but steadily. The transparency increased slightly as it always does after a rise of the water level. The conductivity dropped to lower values. The temperature increased irregularly and in July and August temperatures as high as 34°C were recorded.

During this wet period the temperature in the deeper water remained unaffected. Warming up was restricted to the surface. Oxygen on the average increased slowly and to some extent followed the fluctuations of the temperature. Compared to the previous dry period the fluctuations in oxygen content were smaller even in the daily observations.

This apparently favoured the development of crustaceans, for *Bosmina*, *Ceriodaphnia cornuta*, *Cyclops*, *Diaphanosoma brachyurum* and *Diaptomus* and also rotifers as *Ascomorpha saltans*, *Asplanchna*, *Conochiloides coenobasis*, *Polyarthra* appeared in great numbers. Most of these organisms showed a pronounced vertical migration in the surface layers, especially *Ascomorpha* and *Conochiloides*. It seemed that the flagellates were somewhat less numerous, at least until July 3–7 and this might also be true for *Dictyosphaerium* and *Closterium*.

Some typical elements in the plankton of this period deserve more detailed attention. As previously mentioned for the normal river, *Eudorina elegans* together with *Diaphanosoma* increased in numbers shortly after the beginning of the rainy period. Here we observed this phenomenon again. *Eudorina elegans* after a maximum development on June 10 disappeared from the plankton whereas *Diaphanosoma* remained approximately at the same level. For crustaceans the circumstances appeared to be favourable. They will probably disappear again in the dry period. This might be sustained by the fact that numbers were decreasing at the end of August.

A few species in the plankton of the open water, present in small numbers, originated from plankton that developed in the drowned forest regions of the lake. An increasing number of dead leafless trees were observed in the Afobaka region about May 27. Larger areas of the water surface were exposed to sunlight resulting in a greater development of filamentous algae (*Spirogyra*, *Mougeotia*). Amongst these algae bottom dwelling organisms were found like the crustaceans *Cyclestheria hislopi*, *Euryalona occidentalis*, *Iliocryptus* and oligochaetes like *Aeolosoma*, *Aulo-*

phorus and *Dero*. Shallow water forms also may be found, like *Daphnia* sp. which appeared at Afobaka in July for the first time.

During July and August the area of the lake increased steadily. The distance from the Afobaka sampling station to the inflow of the river into the lake increased accordingly, finishing the influence of the river plankton on the plankton community at Afobaka. On June 1, brown turbid water from the river resulting from heavy rains reached as far as 5 km south of Afobaka. The colour of the water at Afobaka, however, remained dark and no influence of the river water was observed.

From June 1 onwards the situation at Afobaka was only influenced by rainwater and water from forest areas. Much depends on action of the wind even at low velocities.

The plankton became more and more autochthonous and typical for the lake. This was illustrated by the development of many Chlorophyceae such as *Dictyosphaerium*, unidentified conglomerates of green cells, *Melosira*, Crustacea and Rotatoria. Further observations will be required before an opinion can be formed about the stability of the plankton community.

It was interesting that the rotifer *Sinantherina spinosa* appeared in the plankton until July 22 apparently living in zones with a low oxygen content. During this period thousands of turbellarian worms (*Catenula lemnae*) were found in the water layer just above and below the oxygen limit. In this lake community, the diatom *Melosira granulata* was apparently replacing *Eunotia*.

Station 2. *Kabel* (Figs. 12–13; Table X)

Kabel is located about 12.5 km south of Afobaka where the railroad from Onverwacht via Kwakoegron and Brownsweg ends. In the former village there was a hospital which has been moved to the village of Brownsweg. With a corial, equipped with an outboard motor the distance could be covered in about half an hour, after the water level had risen. In the normal river it used to take several hours to reach Kabel especially in dry periods when the water was shallow. Rapids and hidden rocks considerably impeded the progress of the corials.

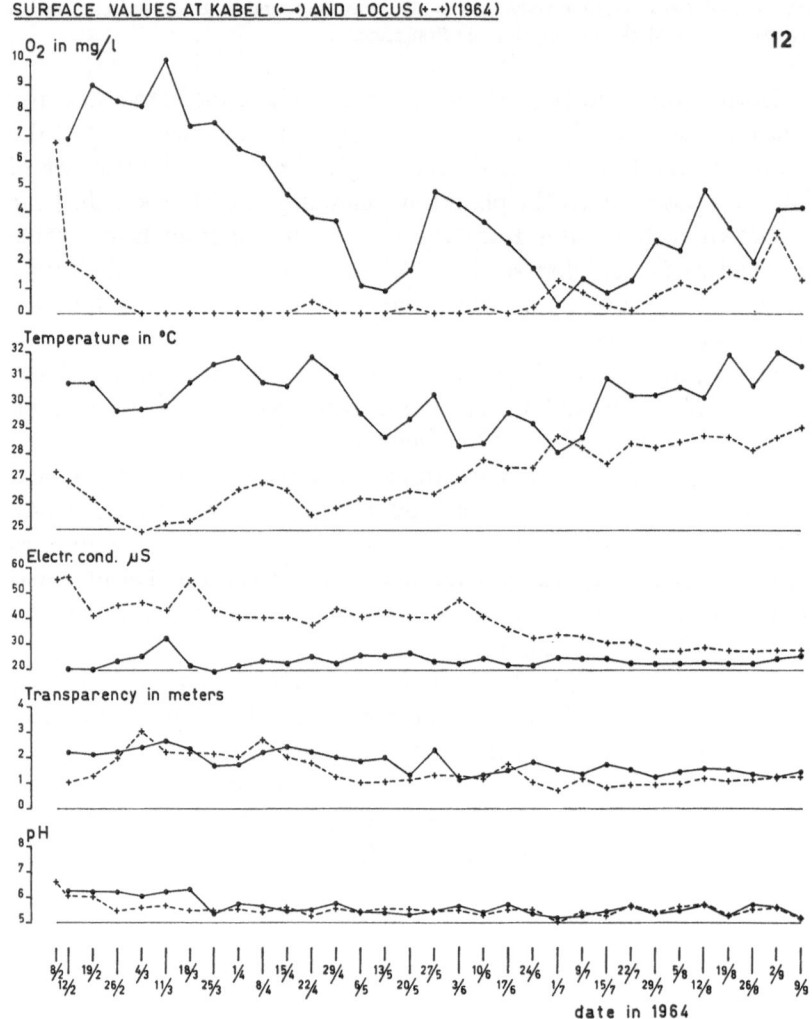

SURFACE VALUES AT KABEL (•—•) AND LOCUS (+--+)(1964)

12

51

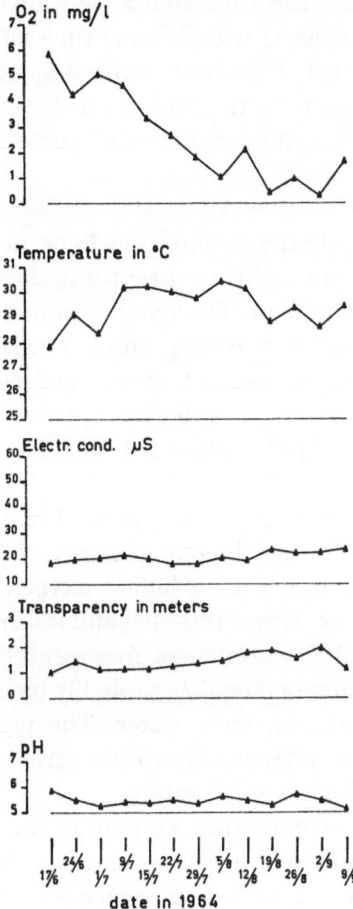

The sampling station was located in the middle of the river just downstream from the northern point of Kapasi island where two branches of the Surinam river join. The samples were taken on the same day and at nearly the same hour as at Afobaka. After the dam at Afobaka had been closed it took some time before the water became stagnant at Kabel. The effect of the stagnation was not quite as sudden as at Afobaka. As the station was located at the junction of two river branches, a more complicated succession of events was recorded.

At the time of the first samplings on February 12 the depth of the river was about 5 m and the current was slow. On February 19 the river was nearly stagnant. The water became saturated and even oversaturated with oxygen during three weeks. From March 25 until May 13 the oxygen content of the surface water decreased gradually. The first anaerobic situation appeared near the bottom in the week preceding April 8. Dead fish were not observed near the station.

The transition period at Kabel from March 25 until May 13 covers a dry season. During the following period from May 14 until August 12 re-stabilization with a higher oxygen content occurred in the superficial layers. It was retarded and interrupted by a period of rain, which affected all waterlayers. As a result the oxygen content was raised rather suddenly (May 27), probably by horizontal currents originating from upstream river water. The period from May 27 until July 1 represents a second transition period, this time at once followed by a period of re-stabilization.

When considering temperature we found that in the first weeks recordings were comparable to values found at Afobaka before the stagnation. Until April 1 we see an increase in temperature in all water layers. Then with the water level rising rapidly, a decrease in temperature occurred during two weeks, followed again by an increase affecting the superficial water layers only. The changes in temperature can be related with periods of stagnation and periods of total mixing of the water layers resulting from the seasonal changes.

The conductivity was generally lower than at Afobaka. The gradual increase from February 19 until March 11 was only recorded in the surface layer and coincides with very high oxygen values. The rainy season again caused a dilution. The variations in the values on the whole were not so great as at Afobaka and during August the conductivity was remarkably constant.

As at Afobaka the pH dropped to 5.5 right at the beginning of the period of transition.

The transparency of the surface layers increased (sedimentation), but decreased when the transition period began (plankton development). A steep but shortlived decrease and increase was found when the wet season started and the water level rose quickly (May 20–27). Finally the values were equal to those found at Afobaka in the first weeks of September. In general the transparency was higher than at Afobaka which might be due to the larger water volume at the sampling station and a difference in the development of plankton.

Observations of plankton may help to distinguish between seasonal changes.

The increase of crustaceans in the plankton began several weeks after the stagnation of the water (April 8 and February 19 resp.) and was accompanied with the appearance of many flagellates. This period lasted from April 8 until May 27 when the rainy season started. The species were the same as at Afobaka with *Bosmina, Ceriodaphnia, Cyclops, Diaphanosoma brachyurum, Moina micrura, Eudorina elegans, Euglena, Peridinium, Strombomonas ensifera, Trachelomonas* spp. Unspecified unicellular flagellates appeared also.

It seems that the crustaceans found favourable conditions some weeks after the stagnation of the water, when the permanent oxygen-free zone had stabilized (April 8–May 27). If we look more closely at the relative abundance we find large numbers of *Eudorina* and *Bosmina* showing some increase from April 1–29. This was also observed at Afobaka in the transition period though here *Diaphanosoma* and not *Bosmina* appeared together with *Eudorina*.

The biocommunity disappeared in the period from May 27 until about July 1. In this period of heavy rain strong currents disturbed the total water volume. The crustaceans reappeared first, on July 1, but the flagellates were not observed until August 12. The period with a permanent oxygen-free zone began on July 1 and it probably took some time before the motile flagellates developed.

In the weeks from July 29 until August 26 *Eudorina elegans* developed strongly. *Eunotia, Rhizosolenia, Melosira*, desmids and Protozoa developed also. Many elements of the plankton community from upriver were evidently transported by horizontal currents during the previous period of rain. Later on the species *Melosira granulata, Dictyosphaerium*, several flagellates and crustaceans, and the rotifer *Conochiloides coenobasis* remained part of the plankton community appearing regularly and sometimes in large numbers. This biocommunity is almost similar to that of Afobaka.

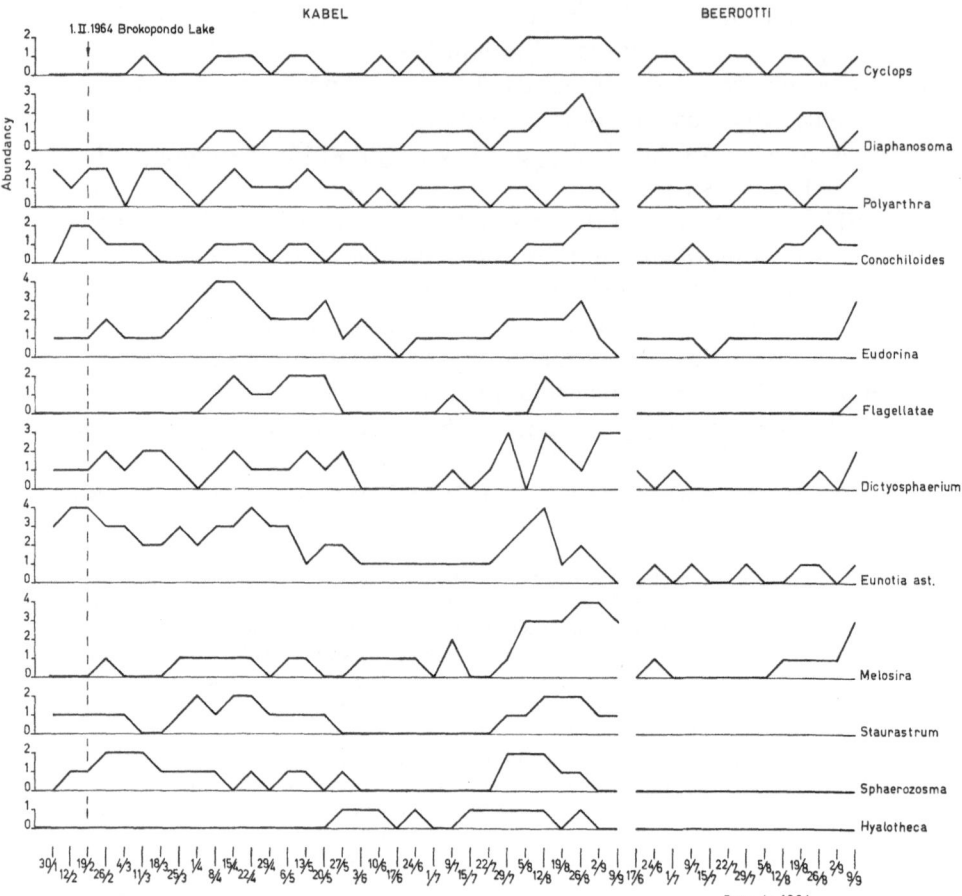

Several plankton organisms originated from the drowned forest habitat. At Afobaka they were found in the open water especially in July and August. *Cyclestheria hislopi, Euryalona occidentalis* and *Iliocryptus* already have been mentioned. At Kabel we found *Alonella, Chydorus, Moinodaphnia* in addition. Of the rotifers *Sinantherina* and *Filinia longiseta* were mentioned. Filamentous algae (*Spirogyra, Mougeotia*) were found regularly and also threads of *Hyalotheca* and other desmids like *Cosmocladium, Micrasterias arcuata* and *Micrasterias radiata* f.*brasiliensis*.

At the end of August colonies of *Rhipidodendron huxleyi* could often be found. This organism appeared during the rainy season in the river, probably transported from tributaries (see Pokigron). Its appearance at Kabel indicates that what happened upstream influenced the situation at Kabel. In this connection the diluting effect of rains during July and August is understandable. It resulted in a low conductivity during July and August at Kabel, thus favouring the development of a special plankton community found in that period. The decrease of plankton in the period from May 27 until about July 1, however, was caused by currents and not primarily by dilution.

Other organisms which were typical of the running river disappeared when the water became stagnant. At Kabel these were ephemerids from rapids and bottom, eggs of dragonflies, pieces of Podostemaceae, veliger larvae and also spiculae of sponges.

Station 5. *Beerdotti* (Figs. 12–13; Table XI)

The sampling station was located south of the little bush negro village of Beerdotti; roughly in the centre of the future lake at about 12.5 km from Kabel. Sampling started on June 17 some time after the water became stagnant. The water level rose about 7 m. During the weeks before July 22 and August 19 the rise was fast.

The temperature was several degrees lower than at Afobaka and about equal to the values at Kabel. After August 19 the temperature dropped, whereas at Kabel the temperature rose.

It should be mentioned that the trees near Beerdotti had not yet died and were still green, while those near Kabel had been dead for a long time. As a result near Kabel the sun could reach the water better. Moreover the influence of colder river water from the south was noticeable in the week before August 19. Finally, warming up was restricted to a very thin layer at the surface probably because at Beerdotti the water was more sheltered against wind. Total mixing of the water layers was observed on July 1 resulting in uniform conditions from top to bottom.

Sampling was started sometime after stagnation. The oxygen value at the surface had already dropped. It gradually decreased until August 5, when low oxygen values were found in a surface layer of 2.5 m. This progress was interrupted only briefly on August 12.

The conductivity was lower than at Kabel and Afobaka. In the week before July 22 especially, the deeper water layers were influenced by the influx of cold, fresh water. Thereafter a gradual increase was recorded up to a level equal to that at Kabel and Afobaka.

The pH was about 5.5, temporarily decreasing in periods of rain.

The transparency of the water increased slowly from 1 to 2 m. No heavy or continuous rains fell in July and August, and short showers had little effect. The value found on the last day of observation, September 9, resembles those found at the other stations after stabilization.

Plankton was practically absent during the period of observation, as the river contained virtually pure rain water in these weeks. On September 9 *Melosira, Eudorina* and other organisms showed the first signs of development.

Station *Sara* (Figs. 10, 14–15; Table XIII)

This station was located in the former Sara Kreek, a tributary of the Suriname River. It joined the river near Koffiekamp. The stream meandered through overhanging trees and the current was sluggish. Navigation was impeded by lack of depth. In dry times the depth in many places was no more than 50 cm and several dead trees lay across the stream. The sampling station Sara was chosen at a place that could be reached within one hour during the dry period.

This station was interesting for several reasons. The stream differed in character from the Suriname River. The turbidity was high and the colour brown owing to iron and detritus. The station was located among trees, unlike stations on the Suriname River.

The samples at station Sara were taken on the same days as at Afobaka, Kabel and Beerdotti, but in the morning. Sampling started on the same date as at Afobaka. The observations in November, December and January are therefore from the original stream. After February 1, it took only two or three days until the water at Sara became stagnant.

For comparison with the data found at Afobaka see Fig. 10. Fig. 14 also shows values found near the bottom.

Sara Kreek before stagnation of the water

The results of the measurements in the original Sara Kreek indi-

SURFACE AND BOTTOM VALUES AT SARA AND LOCUS (1964)   **14**

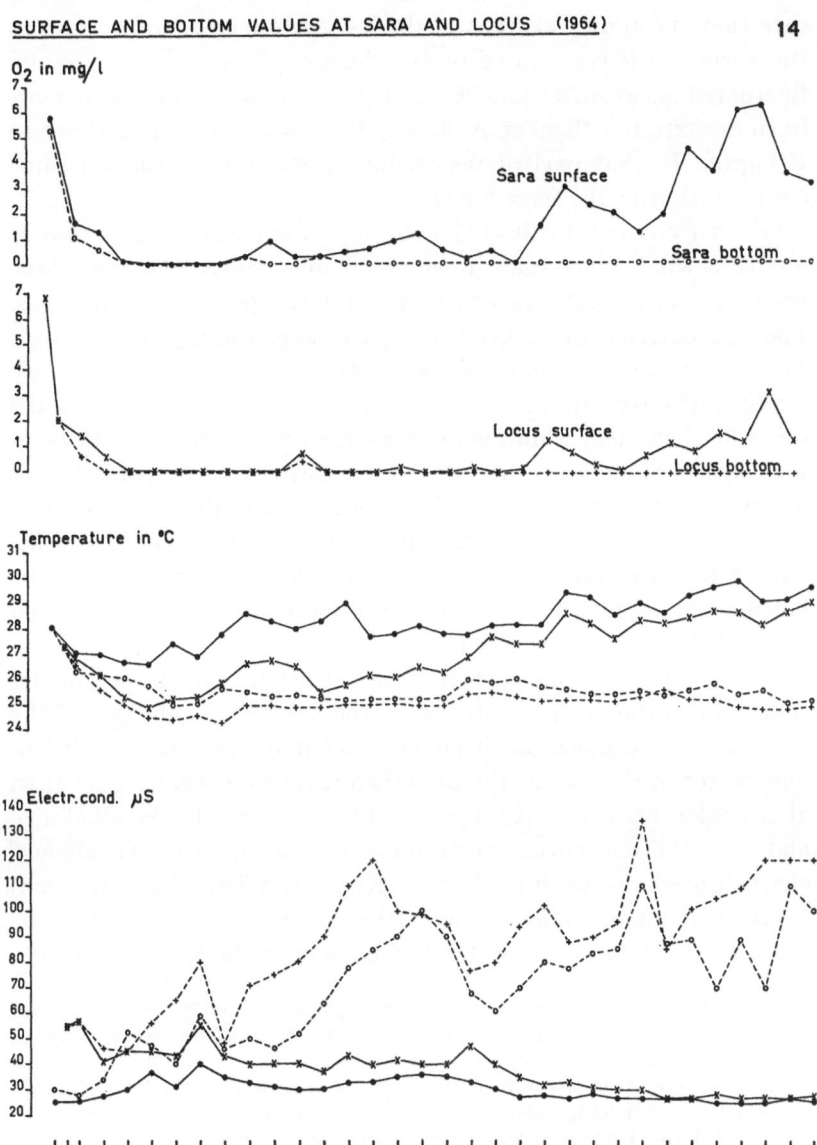

cate that the temperature was always some degrees lower than in the Suriname River, caused by the shadow of the trees. The values fluctuated between 25° and 28°C. These fluctuations were apparently more extreme than at Afobaka; they were more like those at Pokigron. In that part of the Suriname River the situation came closer to that in the Sara Kreek.

Oxygen content fluctuated between 5 mg/l and 7 mg/l, thus it was somewhat lower than in the Suriname River. This may have been caused by the sluggish current and the large amount of detritus. The conductivity was high in comparison to the Suriname River. The concentration of minerals was higher, as the stream was very shallow. The rivulets gave the same picture (Chapter 2). When the water level was high, dilution by rainwater became apparent through a decrease of conductivity and temperature. The turbidity, as measured with the Secchi disc, however, was not affected. This may be due to the fact that the water level rose about 4 m in the narrow stream bed, causing a considerable turbulence. Detritus from the bottom was carried downstream and organic matter from drowned parts of the forest was added.

As the volume of the water flowing in the Sara Kreek was smaller than that in the Suriname River, in the Sara Kreek changes of the waterlevel took place much faster. Except for the short period of high water in December, the pH was always somewhat higher than at Afobaka, because this water was better buffered. The alkalinity and also the concentration of minerals was higher. We already characterized the water of the Sara Kreek as a Turbid Brown (iron) Water and microscopical examination showed iron bacteria (type *Leptothrix*). The iron content is higher than in the Suriname River.

The development of plankton in the Sara Kreek was poor and it differed from that of the Suriname River. Ironbacteria, fungi imperfecti (*Actinospora*), *Euglena*, *Trachelomonas caudata*, *Surirella*, *Synedra*, *Closterium*, Rhizopoda and Rotifera are found regularly. Also mosquito larvae, eggs of dragonflies, nematods, *Cyclops*, statoblasts of *Plumatella*, cercaria and larvae of ephemerids. *Rhipidodendron huxleyi* was found during high water and *Oscillatoria* threads were more numerous than in the Suriname River. The plankton contained many bottom dwelling organisms, which tallied with the slight of the water, and also elements from stagnant parts of the stream. The influence of the slow current was apparent from the regular occurrence of *Cyclops* and motile flagellates. Most characteristic of the plankton, however, were the iron bacteria, fungi imperfecti, Rhizopoda and filamentous blue algae. The fungi are interesting as they live in water with decaying organic matter and low light intensity as was the case in the Sara Kreek. Many dead trees and masses of decaying leaves were found along the banks and in the small rivulets in the wood.

The filamentous blue-green algae may respond to the same factors in the environment. They may have originated also from the roots of the waterfern *Ceratopteris pteridioides* which was found floating in a swamp upstream and from other stagnant pools. Iron and iron bacteria may have originated from water welling up from the streambed. Wells with a high iron content were found in the dry riverbed near Brokopondo in the Suriname River. It appeared, however, that these were more frequent in the Sara Kreek region. Another tributary of the Suriname River, the Grankreek, also showed high concentration of iron illustrated by the presence of many iron bacteria. The elements of the plankton community characteristic of the Sara Kreek and Grankreek were hardly ever found in the Suriname River.

Sara Kreek after stagnation of the water

After the closing of the dam we observed the same rise in water level as at Afobaka, however, somewhat retarded because of the distance (Fig. 10). The oxygen content dropped more sharply than at Afobaka. This was caused by the more sheltered situation resulting in a more complete stagnation. The reduction processes were more active in the water amongst the trees. These factors combined to prevent a temporary oversaturation with oxygen just after stagnation, as recorded at the stations Afobaka and Kabel. The paucity of plankton in the Sara Kreek and especially the nearly complete lack of green organisms must be taken into account too. After 5 weeks the water at the station was completely anaerobic from top to bottom.

The situation remained like that for about 4 weeks, until March 25, after which date an increase of oxygen was found. At the same time a great number of unicellar motile flagellates appeared at the surface. The increase in oxygen remained limited to 1 mg/l and was restricted to a very thin water layer at the surface. The same phenomenon happened at that time at Afobaka.

The lower oxygen content after the stagnation in the Sara Kreek, when compared to Afobaka, was a result of a very severe stagnation and a strong reduction. At the surface a thick brown film of iron bacteria, mixed with tiny turbellarians (*Microstomum*) and protozoans developed soon after the impoundment of the water. The film covered the whole surface, especially in sheltered spots amongst

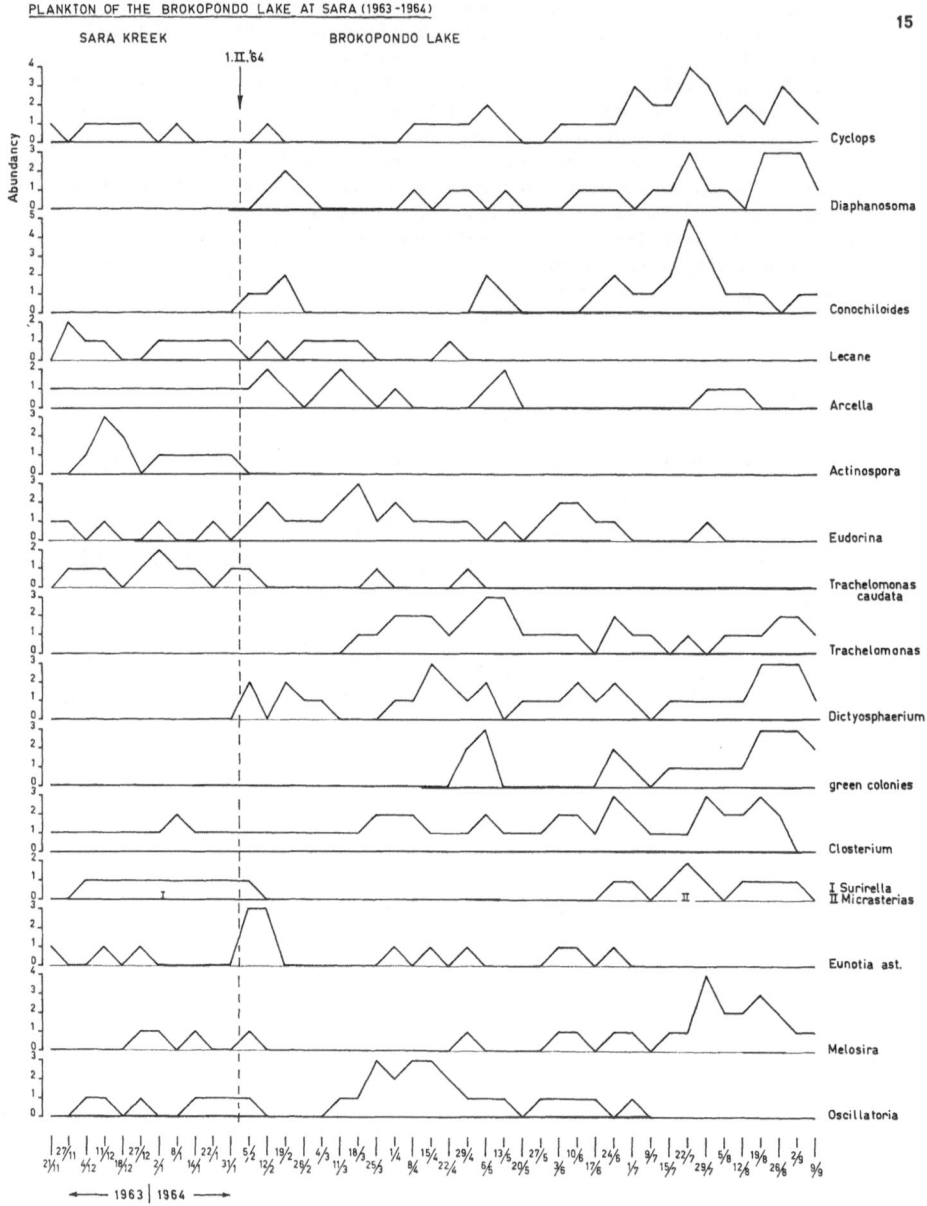

PLANKTON OF THE BROKOPONDO LAKE AT SARA (1963-1964)

15

SARA KREEK   BROKOPONDO LAKE

1.II.'64

Abundancy

Cyclops

Diaphanosoma

Conochiloides

Lecane

Arcella

Actinospora

Eudorina

Trachelomonas caudata

Trachelomonas

Dictyosphaerium

green colonies

Closterium

I Surirella
II Micrasterias

Eunotia ast.

Melosira

Oscillatoria

← 1963 | 1964 →

trees. It disappeared in the rainy period when heavy showers disturbed the surface. The colour of the water changed and, especially near
the bottom, a brown colour was found due to dissolved humic substances. A smell of H₂S was sometimes noticeable. In the brown film
bubbles of gas appeared. The nature of this gas has not been analyzed
but it was supposed to be methane. A low pH is, as a rule, not
favourable for the growth of methane bacteria; the disturbance of
the environment, however, may have given them a temporary chance
to develop.

In the week before April 15 oxygen decreased to 0.3 mg/l. A
change must have taken place in the water as most of the zooplankton organisms disappeared and the transparency diminished.
After April 15 the oxygen content gradually increased again to
about 1 mg/l. At this period the rains ceased and the temporary
stationary situation probably stabilized the biocommunity. At the
beginning of the rainy season on May 20, the oxygen content diminished again, right down to zero on June 17. The rain must have
caused considerable disturbance. Oxygen was found near the surface
and this limited supply was exhausted very quickly after mixing
with anaerobic water. The plankton community was also affected.

As the level of the water rose steadily, the area of open water increased when the trees gradually disappeared below the surface. The
conditions became similar to those found at Afobaka station and
finally June 17 the oxygen content reached values comparable to
those at Afobaka. The composition of plankton, the conductivity,
the transparency and the pH were also comparable to those at Afobaka, but the oxygen values and the temperature remained lower,
owing to the effect of a very dense vegetation at Sara.

After stagnation the temperature fluctuated around 26°C. In the
dry period it rose to 29°C and after that it was about 28°C. After
June 24 it increased to 29.5°C. In this last period the influence of
greater exposure to sunlight was apparent. The effect of the wind
from different directions increased, causing temporary variations.

The most spectular change concerns the conductivity. Within a
week after stagnation it dropped to the same level as at Afobaka.
Variations occurred approximately parallel to those at Afobaka

though sometimes less abruptly. Transparency increased to a depth of 3 m. This was deeper than at Afobaka and the reason for this could be lack of movement of the water among the trees resulting in a better sedimentation of suspended matter. The pH dropped to approximately 5.5.

The plankton community of the flowing Sara Kreek has been described previously (see also Table XIII). Soon after the stagnation bottom organisms disappeared, and also fungi imperfecti, diatoms and typical species including *Trachelomonas caudata*. The diatom *Eunotia asterionelloides*, seldom found in the Sara Kreek now appeared in large numbers but only in the weeks of February 5 and 19. This was caused by a swift rise of the water level with penetration of water and plankton from the Suriname River. There was a temporary abundance of crustaceans and rotifers which originated both from river plankton and from shallow parts of the stream. On March 11 *Eudorina elegans* increased in numbers. This may be related to the transitional period as at Afobaka.

Three weeks after the dam was closed a dense growth of filamentous algae was found attached to the green leaves of the still living trees under water. The *Spirogyra* threads produced oxygen and large numbers of crustaceans like *Moina, Ceriodaphnia, Diaptomus* and *Diaphanosoma* were found swimming just around the algae. As oxygen was lacking in open water and especially in deeper layers, it was the only place were they could survive. Also young fish were attracted, both by the oxygen and by the crustaceans. The leaves of the trees were covered by hundreds of muddy tubes made by Oligochaeta (*Dero, Aulophorus, Aeolosoma*). Later on when the water level had risen the sessile *Spirogyra* died from lack of light but new growth became possible at a higher level. In the dry period of April and May few floating mats of filamentous algae (*Mougeotia, Spirogyra*) were found producing a considerable amount of oxygen (9.6 mg/l). Oxygen was absent in the water just underneath and around this vegetation. Numerous bubbles of gaseous oxygen were formed and kept the algae floating.

In some places in the former Sara Kreek duckweeds (*Lemna valdiviana, Spirodela biperforata*) began to develop. When the duckweed started to cover the water completely the filamentous algae died of lack of light. The floating algae and duckweeds were crowded with organisms from stagnant water. They probably originated from forest pools which were isolated before the water level had risen, e.g. organisms like *Chydorus, Cyclestheria hislopi*, larvae of ephemerids, dragonflies, *Cyclops, Hydra*, Ostracoda, Hydrachnellae, mosquito larvae (*Culex, Ceratopogon*), Gastrotricha, Bryozoa, Rotifera, Oligochaeta like *Pristina longiseta*, Turbellaria like *Catenula lemnae, Microstomum*. They were also found in the open water, at least at the surface.

The plankton in the former Sara Kreek was restricted to the surface through lack of oxygen. The superficial samples collected with a bucket therefore contained more specimens than hauls with a plankton net. The table deals only with the results of the net hauls.

At the end of March many flagellates developed and at the same time oxygen content increased near the surface. The species collected were different from those occurring in the former Sara Kreek. They were forms commonly found in stagnant water. The total composition was very similar to that at Afobaka except for the

63

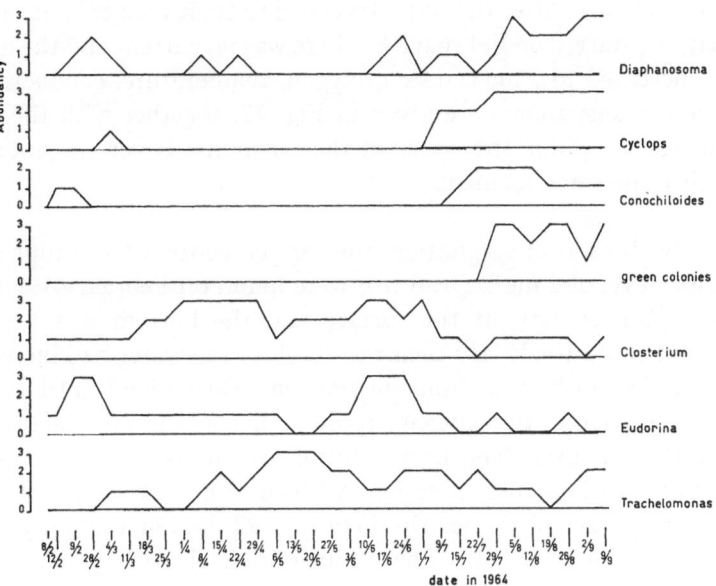

filamentous blue algae, the crustaceans and rotifers. Of the rotifers only the colonies of *Sinantherina spinosa* were numerous. They seem to do well in water low in oxygen content. During the wet season in June and July the plankton was less abundant. In July, however, *Cyclops, Diaphanosoma, Diaptomus* and *Conochiloides coenobasis* increased in numbers. They preceded the development of large numbers of Chlorophyceae, desmids and *Melosira*. It is possible that the motile Crustacea and Rotifera penetrated the Sara Kreek region more rapidly than the more passive phytoplankton organisms, supposing at least that the plankton came from the Afobaka region where the species were already established. The weak currents resulting from the filling up of the lake moved in the general direction of the upper Sara region and might also carry this plankton. The rate of reproduction of the plankton organisms also played an important part in the rate of appearance at the station.

In September the depth of the water at Sara was about 25 m. The rise in the water level followed the same pattern as at Afobaka and gradually the environmental conditions in these two lake regions became very similar.

Station 3, *Locus* (Figs. 12, 14, 16; Table XII)

Station 3 was located at the junction of the Sara Kreek and Locus

Kreek. Before the dam was closed the station could be reached by boat in 3 hours. After the water level had risen, it took only one hour. Sampling started on February 8. There was no current and the depth had increased to 2 m. Data on oxygen, temperature, conductivity, pH and transparency are given in Fig. 12, together with the data from Kabel, where the water of the Suriname River stagnated at roughly the same moment.

A few days after stagnation, the oxygen content had dropped to the low level of 2 mg/l. Dead fish were found on February 19, when the oxygen content at the surface and the bottom was 1.5 and 0.5 mg/l respectively. By then the depth of the water had increased to 6 m. As can be seen from the oxygen values found until July 1, the exhaustion of the oxygen supply at Locus was even more complete than at Sara. The Locus station was more isolated amongst trees which stimulated reduction. A film of iron bacteria appeared at the surface at this time. On February 22 *Lemna valdiviana* was found spreading rapidly. The origin of this floating plant and of other organisms from stagnant water living on it is unknown. Somewhere in the old river system there must have been a pool, swamp or dead branch where a stagnant water community could develop.

The complete absence of oxygen in the water from top to bottom continued for 4 months. In this time organisms were only found in a very thin layer at the surface. Just as at Sara filaments of green algae developed. The Lemnaceae formed a cover over large areas of the former Sara Kreek, but disappeared at the end of May when the rains started. It was not quite clear whether the plants died or were simply dispersed over a larger area between the trees.

Contrary to the sequence of events at Sara, the oxygen content did not improve at the end of March. Flagellates, however, were present. Only after July 1, when depth and surface area became similar to those at Sara (and Afobaka), the oxygen values increased, but they remained at a lower level than at Sara. Probably there were more reductive substances than at the other stations.

The temperature first dropped from 27°C to 25°C, probably as a result of the influx of colder water from upstream and from the gradually disappearing forest. In the following weeks the temperature increased again to 27°C on April 8 and then dropped once more. It is possible that the increase was due to a change in colour of the water. After sedimentation the colour of the water turned from brown to black. It was not clear why the temperature dropped after April 8 as no changes in the weather or in water level took place (dry period until May 20). After April 22 the temperature rose gradually and appeared parallel to that at Sara, although it remained slightly lower because of the more intensive shadow.

Though initially conductivity was higher than at Sara, later on in August and September, it was about the same. Increasing depth and volume of the water created a comparable situation. Conductivity finally remained nearly constant at about 40 µS. When at the end of May the rainy period began and the water level rose rapidly, conductivity decreased on account of dilution. In the dry period of April and May there was little change. Near the bottom the values were higher and increased gradually.

As at all the other stations pH dropped to 5.5 appr. Transparency increased in the first weeks but after April 8 it diminished to 1 m.

Comparison with the curves from Kabel in Fig. 12 shows that after July 1, the values at the two stations became similar. The same happened to the plankton.

The plankton community at Locus did not differ much from that at Sara. As no plankton was collected when the water was normal, it is assumed that the difference was slight at the start. Comparison with the first plankton samples at Sara renders this fairly evident.

In comparison with the Sara station the growth of *Closterium* in the dry periods of April and May was slightly better. The development of *Eudorina elegans* during the first few weeks and at the beginning of June, each time following a swift rise of the water level, was also more pronounced at Locus. Locus is a long way from the Suriname River and as a result the influence of its water is slight. *Eunotia asterionelloides* therefore was seldom found and only during the first weeks. Flagellates were always present. Oxygen, however, was always absent at the surface. After July 1 large numbers of Chlorophyceae and desmids appeared and the oxygen content began to increase at the surface. This was much later than at Sara, and it never reached the same level. This may indicate that either the plankton community was producing less oxygen, or that more reductive substances were present. However, the influence of a lower light intensity and temperature must also be taken into account.

4c. Limnological changes of the environment

The data on depth distribution of temperature, oxygen content, conductivity and pH are plotted in graphs. In one series the depth distribution is given for each sampling separately. In another series the variation at each level during the observation period is given. Both graphs may help to increase our understanding of the processes in action during the filling of the lake.

Station 1. *Afobaka* (Figs. 17–20)

Before the closing of the dam the water was mixed from top to bottom. The temperature showed a slight decrease and the conductivity a slight increase near the bottom. Oxygen was near saturation level everywhere. In the daily observations (Chapter 4a) further details are given. It is necessary to mention once again that in the normal river during high water differences between surface and bottom increased. On December 11 the depth was about 7 m and the following variations between bottom and surface were found: temperature 27.8–29.8°C; oxygen 3.7–6.4 mg/l; conductivity 37–24 µS; pH 6.4–6.7. This seems to prove that even under normal conditions oxygen content near the bottom may be low.

After the closing of the dam the temperature was fairly uniform from top to bottom for several weeks. The water was apparently well mixed and gradually the currents and turbulences subsided. Isotherms show that from March onwards the top layer of 0–3 m became warmer while the deeper water gradually became colder. This development was interrupted during the rainy period, starting on March 11, at a moment when the volume of the lake was still fairly small. The temperature was markedly affected from top to bottom. Later on May 27 and in August, rains had only a mild effect on the top layer of 3 m but deeper water showed a stronger response. Undercurrents may have interfered. Diurnal fluctuations became notable in the top layers but were less distinct in deeper layers as heat absorption was restricted nearly exclusively to the top layer. A thermocline with corresponding epi- meta- and hypolimnion was not found at all or only temporarily during the day. Temperature inversions were recorded, as on May 6, but no permanent thermocline developed in deeper water. The wind action as a downward acting force was subordinate to differences in density, which resulted in density stratification restricted to the top layers.

The appearance of the 27° and 26° isotherms in higher waterlevels in July and August clearly shows that the deep water gradually cooled off. This means also that the water was not stabilized thermically and that heat exchange had not yet reached an equilibrium. The lowest temperature to be reached near the bottom will probably be about 25°C which is also the lowest temperature recorded in tributaries not exposed to sunlight.

Like the temperature, the c o n d u c t i v i t y was fairly uniform from top to bottom during the first weeks after stagnation. The iso-conductiva show that the conductivity soon increased in all water layers, synchronously with the isotherms. This increase was interrupted during the rainy period as a result of dilution. Almost down to the bottom the water became mixed again. The conductivity increased downwards. During the heavy rains of June and July the conductivity increased only slightly down to a few meters above the bottom. By this time the volume of the lake had increased and the amount of trapped rainwater was considerable, with the result that the major part of the water became homogeneous. For this reason the 30 μS isopleth-conductive did not move to higher levels after July 15, but ran approximately parallel to the 28°C isotherm. Both remained at a practically constant level until after August 19, when the gradual cooling of deeper water pushed the 27°C isotherm to the top layers. Then the 30 μS moved swiftly upwards.

The cause of the increase of conductivity was evidently a mounting content of minerals in the water. Daily observations showed that in the course of one day a "wave" of conductivity moved down to the bottom. This explained the maxima recorded at 1.5 m on February 19; at 6.5 m on March 18; at 4.5 m on June 10; at 4.5 m on July 15, etc. These maxima were temporary. The minerals inducing the temporary higher conductivity probably originated from decaying organic material, such as dead plankton and organic material

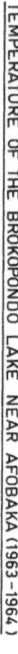

TEMPERATURE OF THE BROKOPONDO LAKE NEAR AFOBAKA (1963-1964)

17

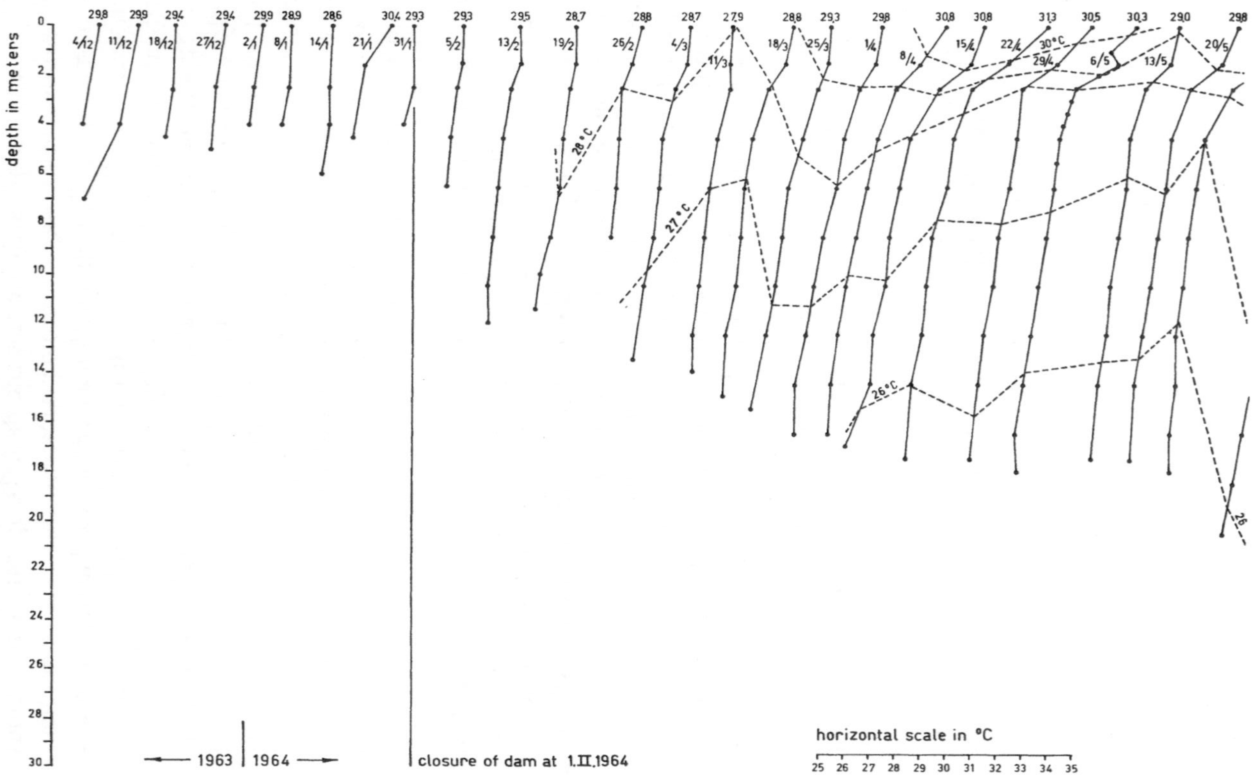
depth in meters
0
2
4
6
8
10
12
14
16
18
20
22
24
26
28
30
28°C
27°C
26°C
30°C
26
← 1963 | 1964 →
closure of dam at 1.II.1964
horizontal scale in °C
25 26 27 28 29 30 31 32 33 34 35

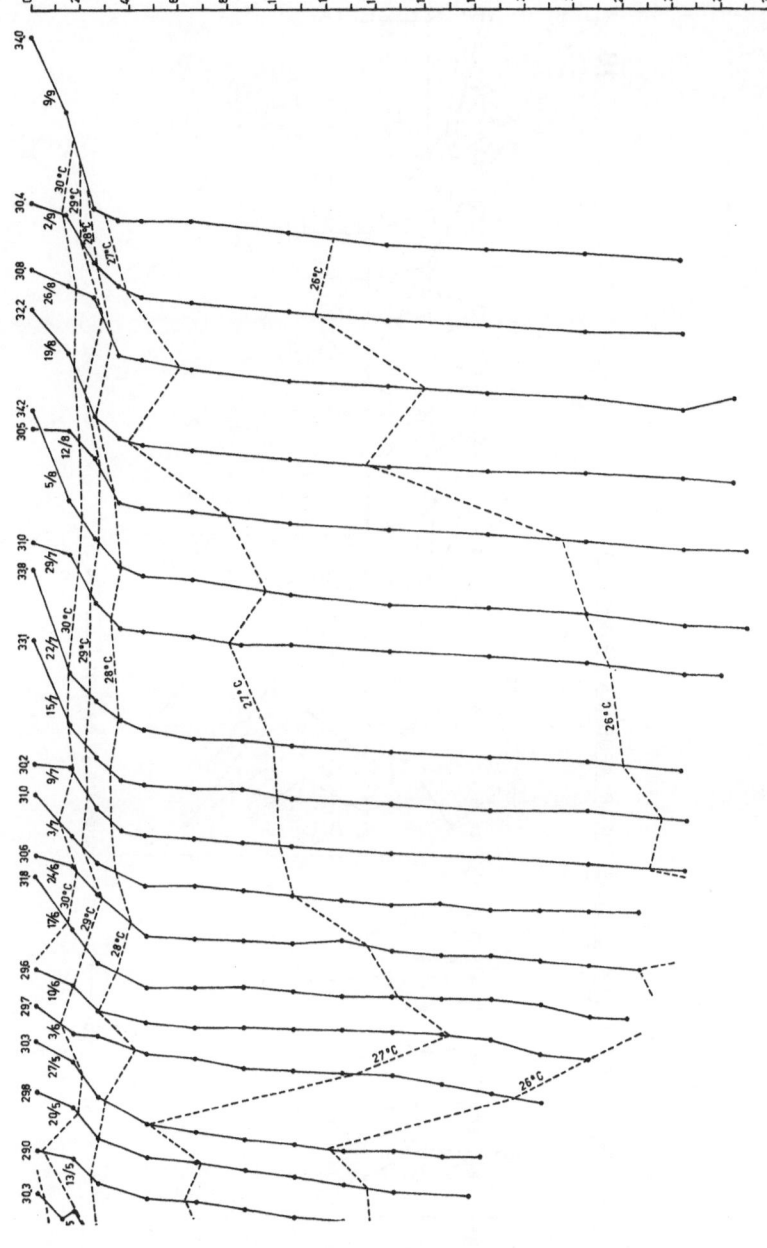

18

OXYGEN CONTENT OF THE BROKOPONDO LAKE NEAR AFOBAKA (1963–1964)

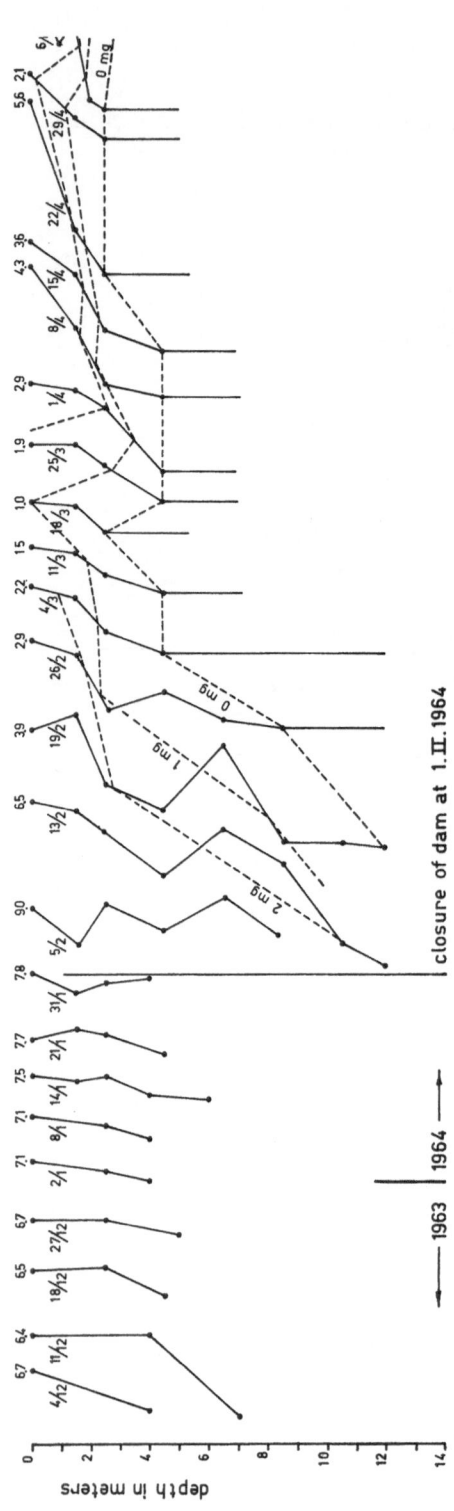

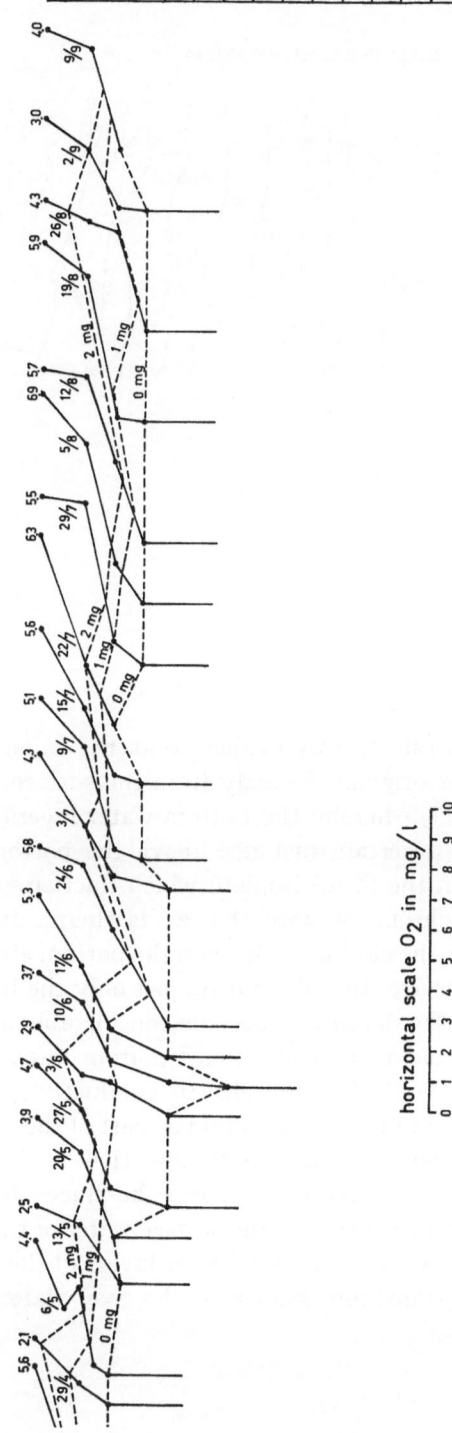

18

horizontal scale O₂ in mg/l

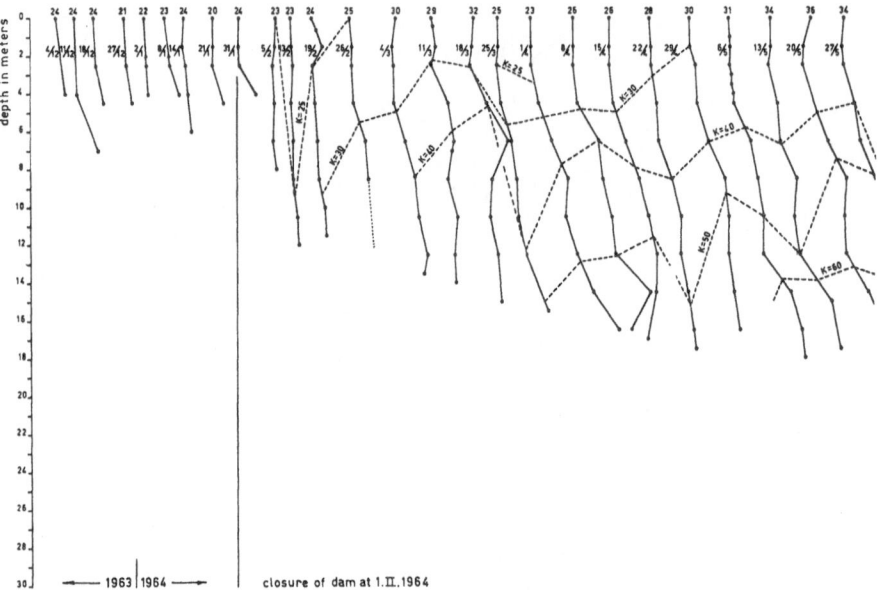

from the forest. It was evident that the mineral enrichment of deeper water originated partly from the surface. There was also a bottom effect, whereby the bottom water layers were enriched by minerals to a certain distance above the bottom. This might be derived from the 50 μS isopleth which fluctuates strongly without showing isochronisms with the 26° isotherm. In shallower water shortly after the closing of the dam the bottom effect was noticeable up to the surface. In July and August only the bottom layers were affected, as the depth had become considerable by then.

Currents and turbulence resulting from rainy periods may also produce "waves" of higher electric conductivity in deep water. A retarding effect may also be found as part of the water flows in from elsewhere moving in a horizontal direction. It should be mentioned that the inflowing river water may be traced by its temperature which is lower than that of the surface water of the lake. This colder water will mix with the cold deeper layers of the lake after passing the zone of transition. Moreover, the river water has a lower con-

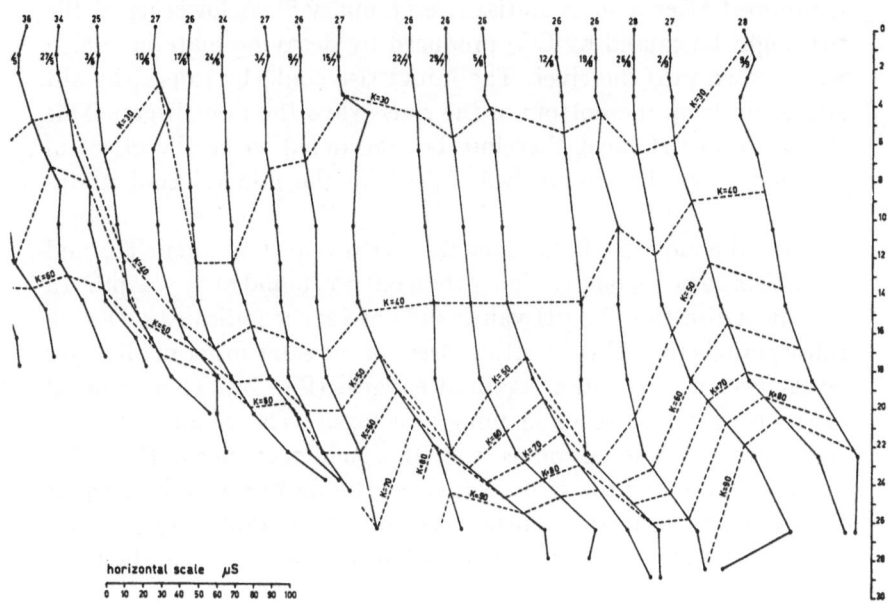

horizontal scale μS

ductivity and it will, therefore, dilute the water with which it mixes. Density differences, due to temperature and to minerals are the main factors in this process.

It is remarkable that in the 0–3 m layer the temperature fluctuations (and also oxygen fluctuations) were very great, but fluctuations in conductivity were negligible. This means, that the chemical situation in the surface layers was more stable than temperature and oxygen. We have seen that some important fluctuations in conductivity did occur but they soon moved downwards. This may be caused by chemical processes at the surface. These produced certain minerals which sank, proving that an increase in density caused by dissolved solids is more active than a simultaneous decrease in density caused by a rise in temperature. Depending on the weather the two factors may work in the same direction resulting occasionally in important fluctuations of conductivity. Considering all this we have to remember that the lake water was not yet stabilized as the final water level had not yet been reached.

In the river the pH was about 6.5. During longer periods of rain it dropped after a short initial rise (January 8). A lowering of the pH might be caused by CO_2 produced by decaying material which was washed into the river. The initial rise could be caused by silt stirred up from the bottom of the river when the rains began. This short rise in pH must therefore be considered to be the effect of turbulence on the stream bed, by which the mineral content increased.

After the closing of the dam the pH dropped to about 5.5 and after February 19, no pH higher than 6.0 was found at any depth. In a vertical direction the pH values did not vary significantly. During rainy periods the pH fluctuations were of the same order in all water levels, becoming constant again afterwards (Fig. 20). This occurred after an initial decrease in mineral content. The sudden increase seemed therefore to be connected with a sudden change in the water at the beginning of the rainy period. As the water was quite deep at that time a bottom effect can be ruled out; so the addition of suspended organic matter from the flooded forest appeared to be the most likely cause.

The water of the running river was nearly saturated with o x y g e n but the concentration decreased slightly towards the bottom. Details of the vertical distribution were collected until one week after the closing of the dam. Shortly afterwards the water was oversaturated for a short time (February 2) and in deep layers maxima occurred which were caused apparently by a diurnal periodicity commented upon in Chapter 4a. In the next three weeks the oxygen content near the bottom decreased rapidly and after March 4 no oxygen was found below 4.5 m. Above 4.5 m the daily fluctuations in oxygen content were very great and ran parallel to the temperature. Temporary inverse stratifications also occurred, e.g. on May 6. In the dry period of April and May the anaerobic zone moved upwards; when the rains started it moved down again. On June 10, oxygen was found to a depth of 6.5 m for a short period. The heavy rains at this time caused a horizontal movement of the water which might have caused a better aeration and also brought down oxygen containing water from upstream.

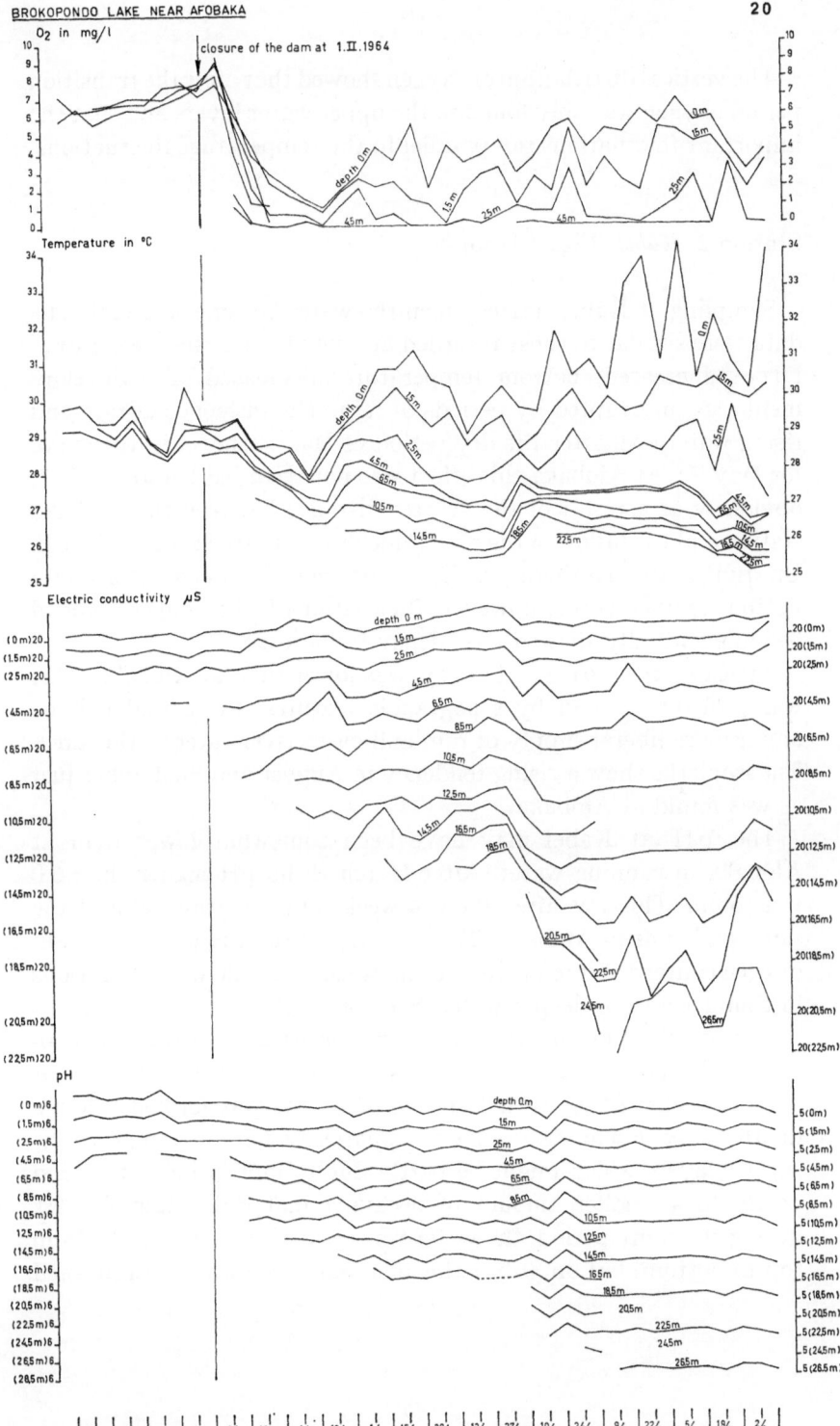

The vertical distribution of oxygen showed that after the transition period oxygen was only found in the upper water layers and that the important fluctuations ran parallel to the temperature fluctuations.

Station 2. *Kabel* (Figs. 21–25)

Sampling at Kabel started when the water became stagnant. The data were similar to those recorded at Afobaka: surface t e m p e r a - t u r e s increased, bottom temperatures decreased. The development was interrupted by periods of rain. The influence of rain and rising waterlevel after the dry period of May is shown in the curve for May 20. At Afobaka this effect was not recorded before May 27, obviously as a result of the greater distance between the Afobaka station and the inflow of the river which was situated south of Kabel. On June 3 and 10 there was little difference between surface and bottom temperatures; however, fluctuations in the upper layer of 0–3 m eventually showed a tendency to increase.

At Kabel the c o n d u c t i v i t y was lower than at Afobaka. This was probably caused by topographic features, the actual volume and a more liberal supply of rain and clean riverwater in this area. The isopleths show a rising tendency in August and September just as was found at Afobaka.

The p H at Kabel may have been somewhat lower than at Afobaka in running water. After March 18 no pH higher than 6.0 was found. This was after about 4 weeks of stagnation: almost the same period as in Afobaka. The lowering of the pH preceded lower oxygen content and temperature and was isochronic with alterations in conductivity in deeper water. See March 18.

In general all events at Kabel were comparable to those at Afobaka. The period of oxygen oversaturation, however, lasted for several weeks. After April 8 an anerobic zone was formed below a depth of 4.5–6.5 m. Oxygen was apparently not restricted to the surface layers which might be partly due to more wind action and partly to a smaller amount of decaying material. After May 20, during the rainy period, the water was aerated for some weeks from top to bottom, but on July 1 the anaerobic zone was re-established

21

TEMPERATURE OF THE BROKOPONDO LAKE NEAR KABEL (1964)

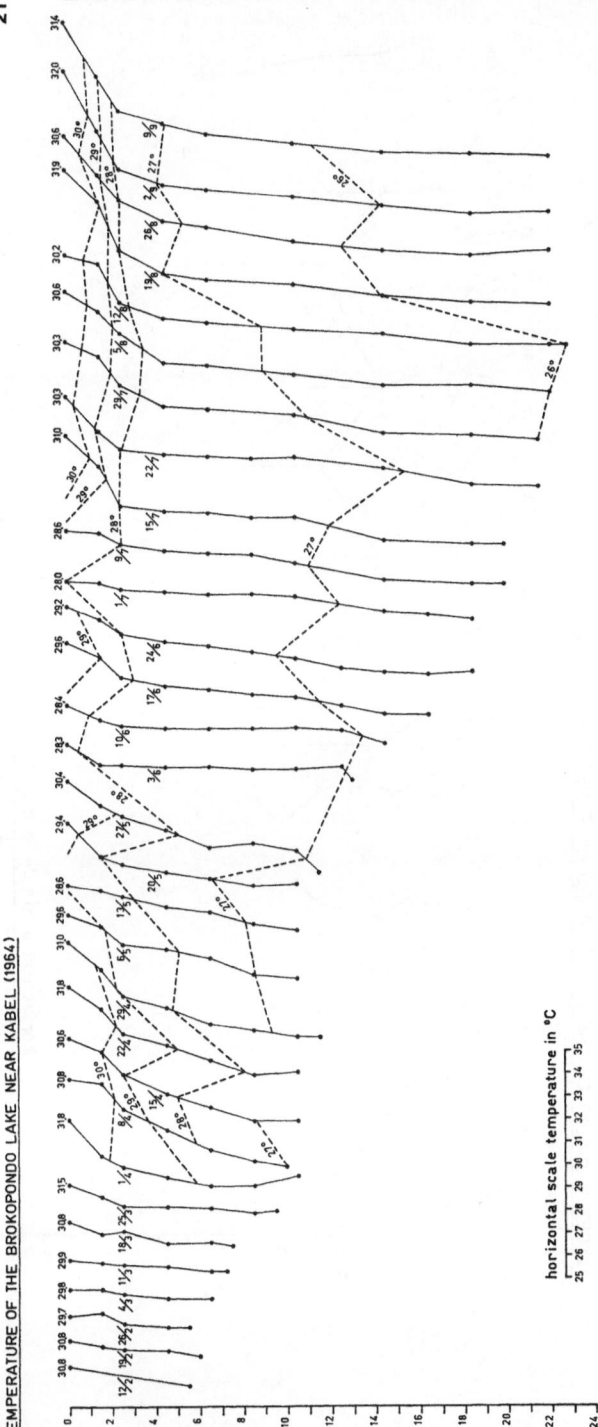

depth in meters

horizontal scale temperature in °C
25 26 27 28 29 30 31 32 33 34 35

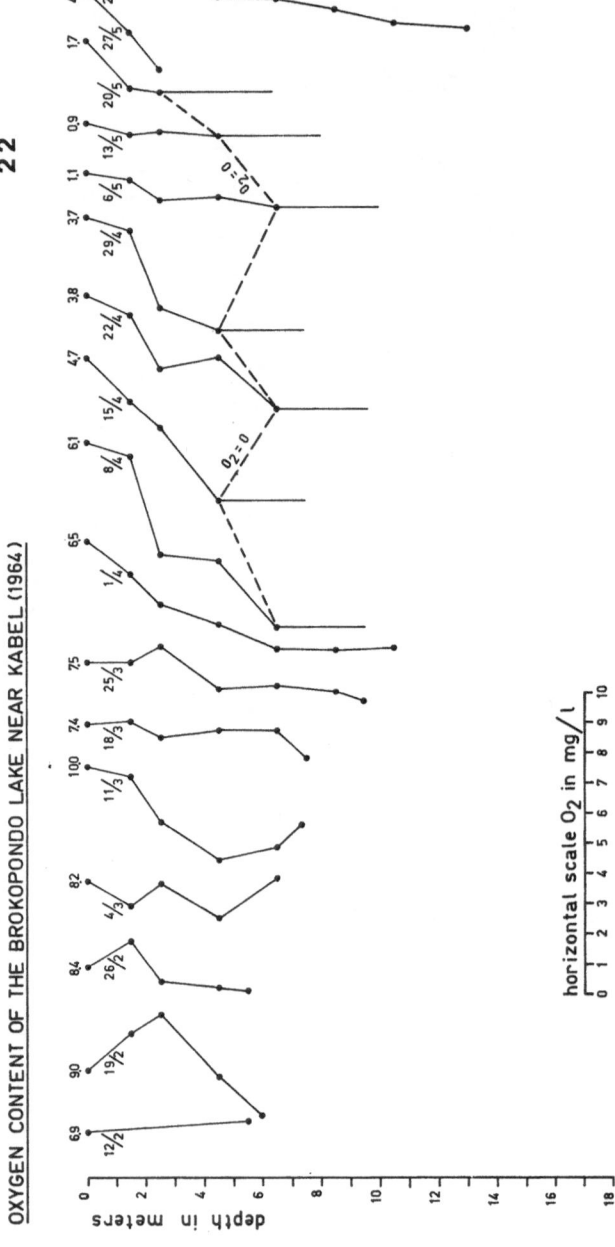

OXYGEN CONTENT OF THE BROKOPONDO LAKE NEAR KABEL (1964)

22

horizontal scale O₂ in mg/l

depth in meters

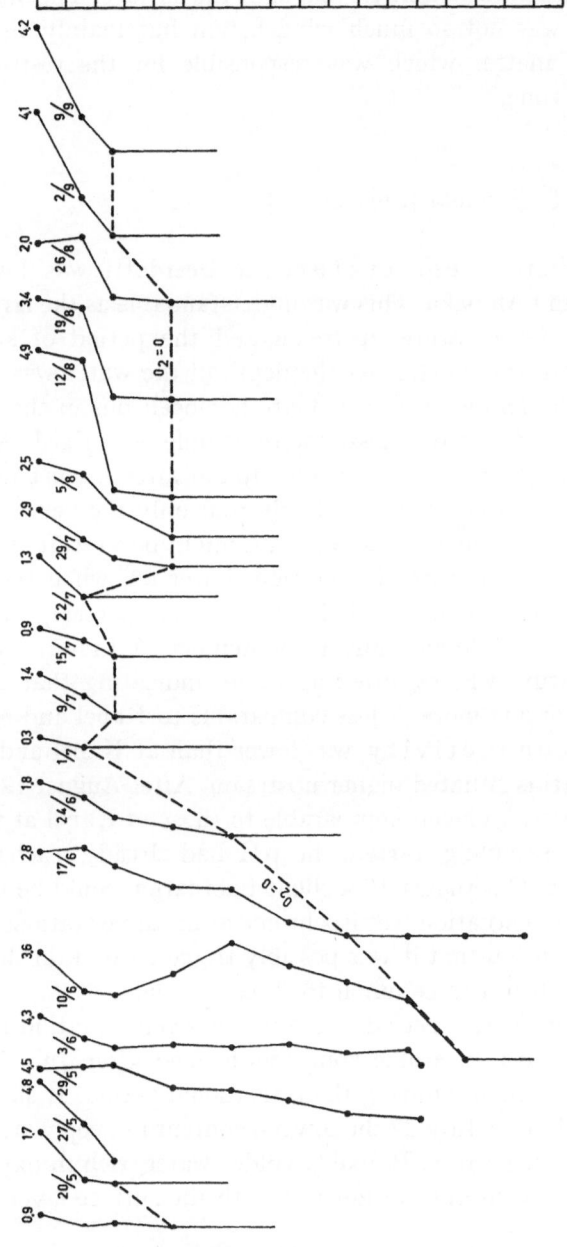

and reached to within 1.5–2.5 m of the surface. This might indicate that it was not so much wind action but mainly the amount of organic matter which was responsible for the restriction of the oxygen zone.

Station 5. *Beerdotti* (Figs. 26–27)

The initial temperature at Beerdotti was lower than at Kabel and Afobaka. This was understandable as the river water had a lower temperature upstream and the period of stagnation at Beerdotti was shorter. As the depth of the water was not yet very great, the temperature was more homogeneous in the whole basin. On July 1 the watermasses were completely mixed. After July 22 the deep water started warming up but after August 12 it was cooling again. Table 27 shows clearly that only the very surface of the water was warmed by the sun. This might be a result of the sheltered situation of the sampling station. Later on, when the water level rose, the surface area and the wind action increased. This occurred at the end of August and in September. A first rise of the water temperature was recorded at 1.5 m, indicating that the situation was becoming more or less comparable to Kabel and Afobaka.

The conductivity was lower than at Kabel and Afobaka as Beerdotti is situated higher upstream. After August 12 the isopleth rose swiftly to levels comparable to those of Kabel at that time.

When sampling started, the pH had already reached a value of about 5.5. On August 19 a slight fluctuation could be noticed. This time the fluctuation was isochronic at all three stations. This favors the conclusion that it was possibly triggered by rain showers, being the only factor in common to three stations.

At Beerdotti a period of oxygen oversaturation has not been recorded. An anaerobic zone was formed after July 29. Sampling took place mainly during the transitional period. It should be mentioned that on July 22 the oxygen content in deep water was higher than at the surface. Probably colder water, rich in oxygen, arrived from upstream but did not mix with the surface layers.

ELECTRIC CONDUCTIVITY OF THE BROKOPONDO LAKE NEAR KABEL (1964)

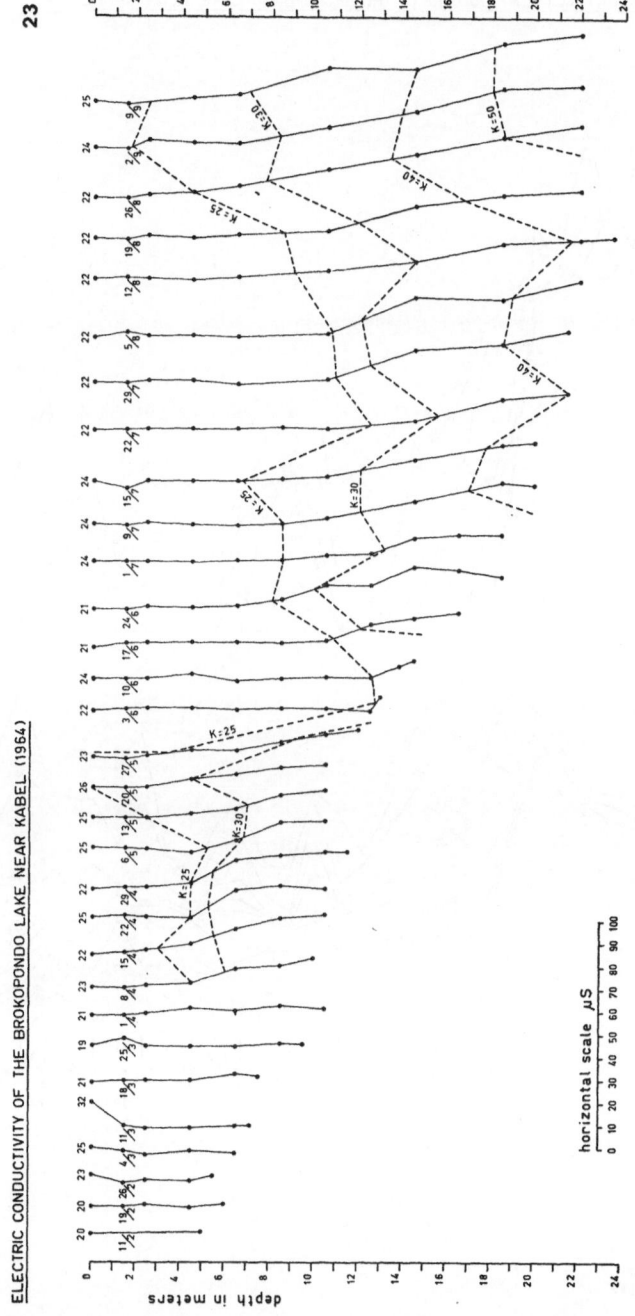

depth in meters

horizontal scale µS

0 10 20 30 40 50 60 70 80 90 100

OXYGEN CONTENT AND TEMPERATURE OF THE BROKOPONDO LAKE NEAR KABEL   **24**

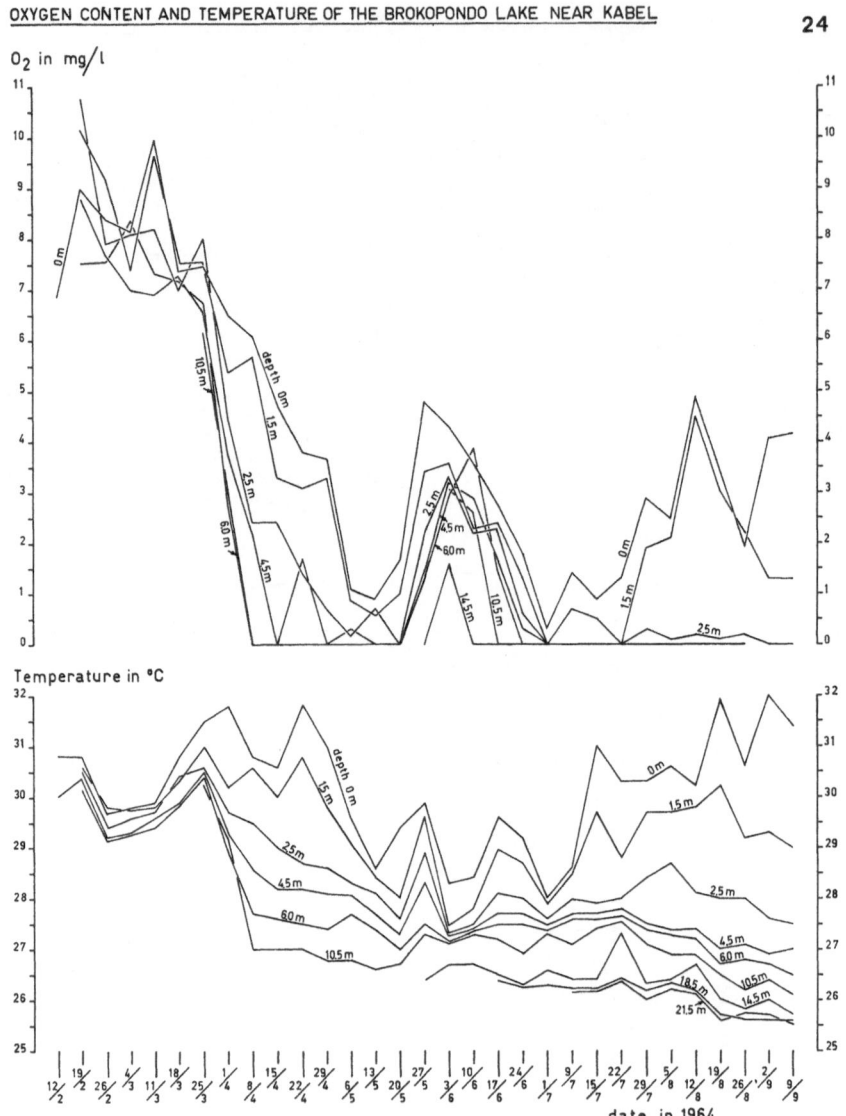

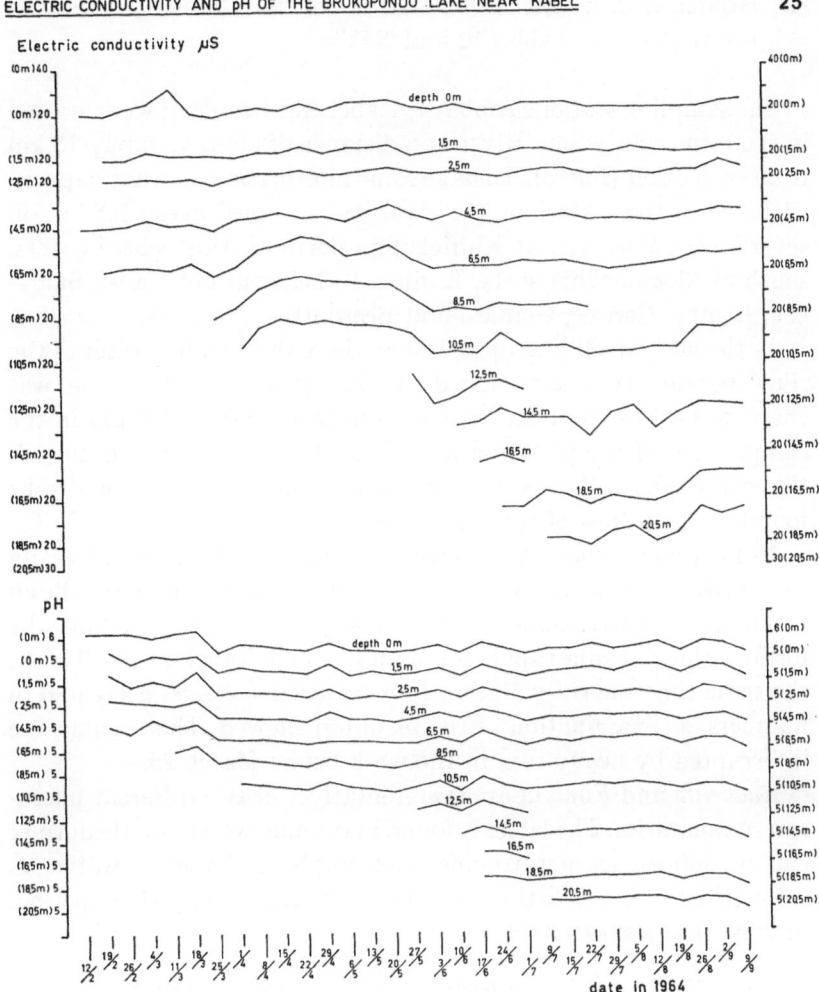

date in 1964

4d. HORIZONTAL AND VERTICAL GRADIENTS FROM AFOBAKA TO BEER-
DOTTI (Fig. 28; Tables 10 and XIV)

The sampling stations Afobaka, Kabel and Beerdotti were located
in the former Suriname River at a distance of approximately 15 km
from each other (Fig. 3). To have some information on what happened between these stations the river was sampled every 2.5 km on
several occasions, viz. at Koffiekamp, north of Aloesoebanja soela,
south of Aloesoebanja soela, Kadjoe, Kabel, south of Kabel, Saidagoe, Lombe, Gansee, Geltoesi and Beerdotti.

As the lake was filling up the zone where the running water of the
river became stagnant moved slowly upstream. This zone was
characterized as the transition zone in which rapid changes in the
composition of the plankton and chemistry of the water occurred.
When the lake will have been filled, this zone will permanently be
found at the inflow of river and creeks.

As mentioned above the transition zone could be followed easily
as numerous *Eudorina elegans* coloured the water green. After March
11 the green wave moved southwards from Afobaka reaching the
former Aloesoebanja rapids on March 18 and Kabel on April 8. At
the same time the river diatom *Eunotia asterionelloides* decreased in
numbers, as examination of the plankton showed. The events were
interrupted by heavy rain in the week before March 25.

Eudorina and *Eunotia* are representatives of two different plankton communities. The latter is found in running water and the former
can be defined as metatrophic. Metatrophy is found in waters in
which the trophic situation has changed in a relatively short period,
or zone (LEENTVAAR 1958).

In Fig. 28 surface values of temperature, oxygen and conductivity at several
stations between Afobaka and Kabel are given. As could be expected during March
the oxygen content increased from Afobaka to Kabel, in April the stretch became
more uniform and when the situation of Afobaka became stabilized at last, lower
concentrations were found once more. A short interruption in this sequence of events
was recorded on March 25 after a few days of rain. The effect of the interruption was
observed also in the temperature gradients and the changes in conductivity and
transparency. In general the transparency decreased in the direction of Kabel and
the highest was found near Aloesoebanja.

On March 28 the vertical distribution of oxygen and temperature was measured
(Fig. 28). The curves for oxygen and temperature measured at the same moment but

BROKOPONDO LAKE NEAR BEERDOTTI (1964)

O₂ in mg/l

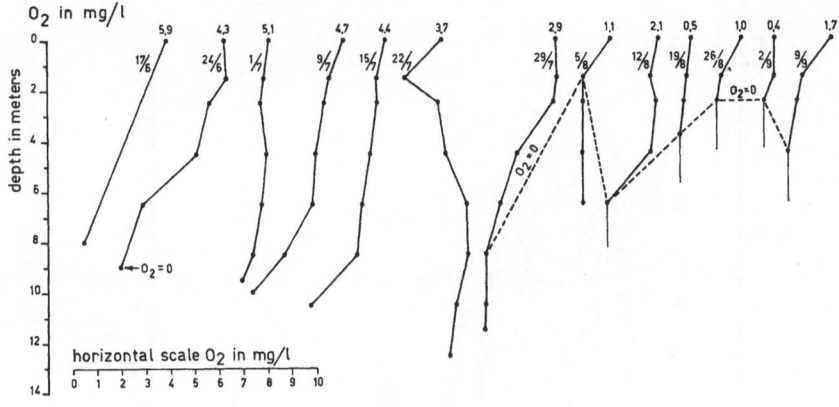

Temperature

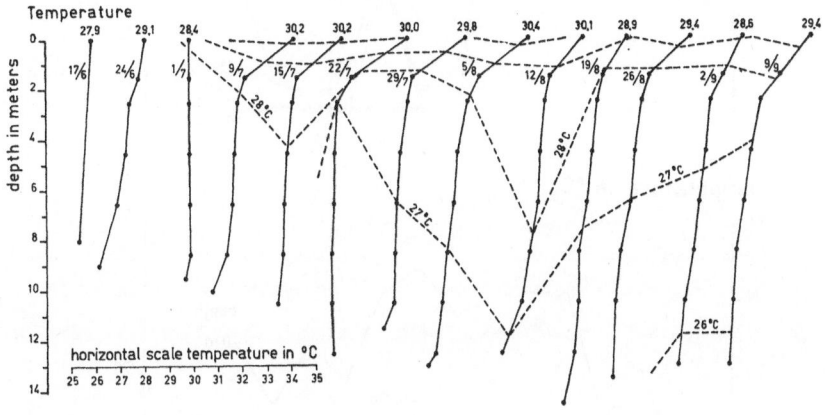

Electric conductivity

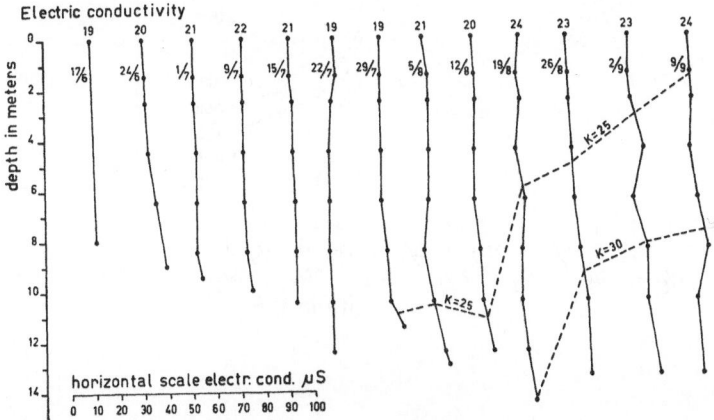

BROKOPONDO LAKE   NEAR BEERDOTTI - 1964

27

O$_2$ in mg/l

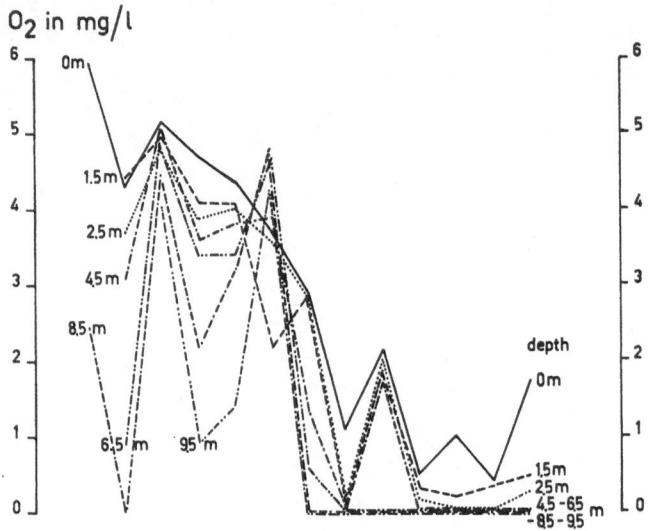

Temperature in °C

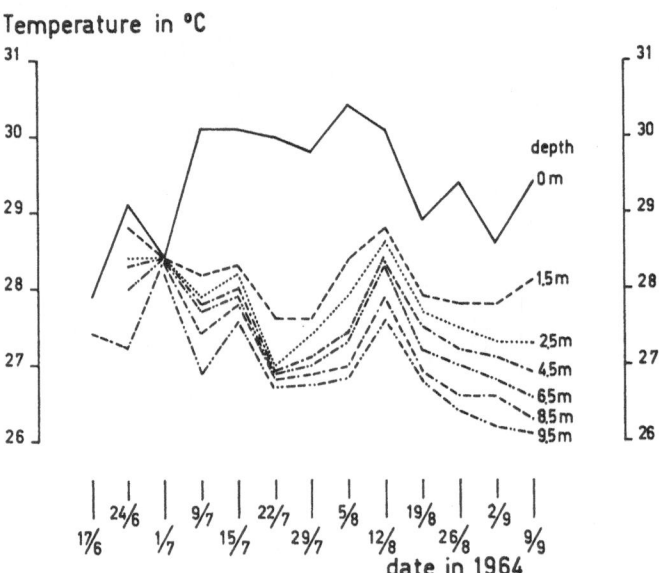

date in 1964

Electric conductivity μS

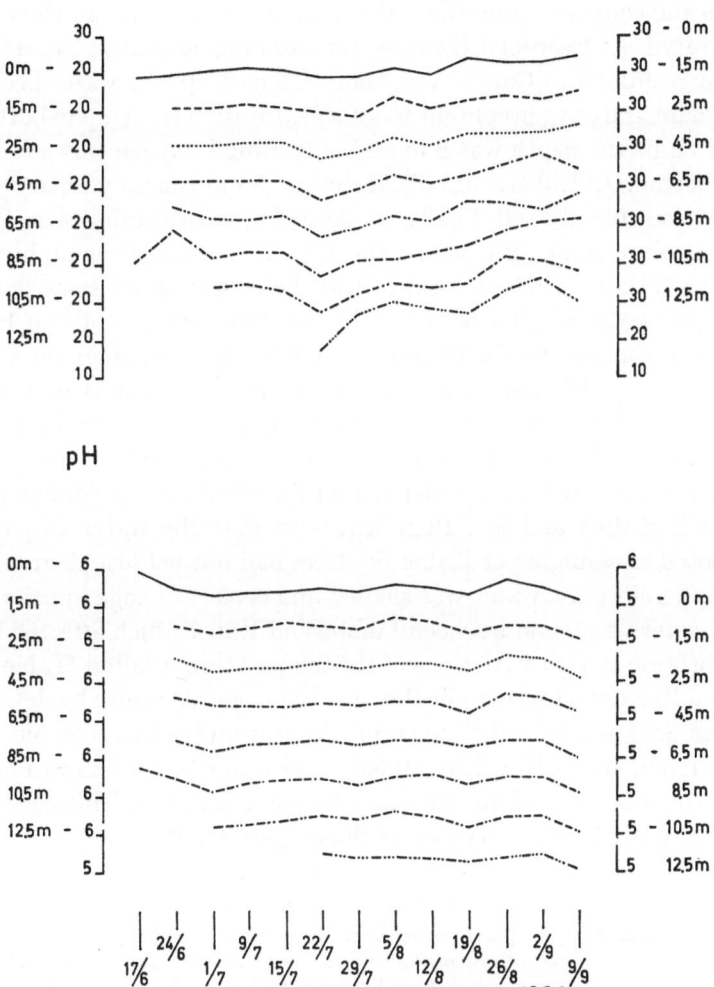

pH

date in 1964

at different stations ran parallel to the curves recorded at one of these stations, in the course of time. This supports the opinion that the same processes were involved whenever the water became stagnant.

On April 29 and 30 a longer stretch of river was sampled. This time the samples were taken from Afobaka right to Beerdotti. The most important plankton organisms are listed in Table XIV. See also Table 10.

In the course of time the differences from one point to the other increased. At Beerdotti the river was running normally; the stretch from Beerdotti to Gansee was about 2.5 m deep, the water became stagnant and oxygen content was near saturation from top to bottom. At Lombé the depth was 8 m and only 3 mg/l oxygen was left near the bottom. At Kabel a stabilized zone with oxygen only until 3–4 m deep was found, and finally at Afobaka, where stabilization had taken place some time before, the oxygen values were again higher.

The variations in temperature near the surface on the same stretch presented a good picture of the different watermasses. From Beerdotti to Gansee the riverwater was fairly cold; the stagnant water between Lombé and Saidagoe was warmer; then a stretch with colder water followed, and it gradually warmed up again near Afobaka. The difference in temperature on the last stretch with stagnant water was caused by a difference in illumination. At Afobaka the trees had died and lost their leaves so that the water was more exposed to sunlight; at Kabel the trees had not yet lost their leaves and therefore the water was shaded and cooler. Change in colour of the water was found between Lombé and Kabel which affected heat absorption so that variations of the temperature resulted (Table 10). From Beerdotti to Saidagoe the river was coloured brown by detritus. From Saidagoe to Kabel green plankton from the metatrophic zone was found; from Kabel to Afobaka the water became brown again but this time the colour was caused by iron bacteria. This changing colour pattern may serve as an illustration for the underlying processes.

The alterations in the plankton communities could also be followed easily (Table XIV). River plankton with *Spongilla* spiculae, veliger larvae, Heliozoa and *Eunotia* was found in deceasing numbers from Beerdotti to Kabel. The passive forms with a higher specific gravity (spiculae, veliger) disappeared sooner from the plankton than others. Lighter organisms like *Eunotia* and Heliozoa were carried over a longer distance but finally disappeared also. In the metatrophic zone *Eudorina* and *Gonium* were found. Finally small flagellates as *Trachelomonas* and *Strombomonas* appeared together with numbers of crustaceans. In the part of the lake between Kabel and Afobaka cladocerans, *Strombomonas* and *Euglena* seemed to find optimal conditions. Near Koffiekamp and Afobaka they decreased in numbers. At these stations, where the water had been stagnant for a long time, extreme conditions prevailed, such as low oxygen content (Chapter 4b and f).

These horizontal gradients give a fairly good picture of the changing situation during the formation of the lake.

4e. DIFFERENCES BETWEEN OPEN WATER AND WATER AMONG TREES (Tables 11–12)

As mentioned already, the temperature in open water was about 2 degrees higher than in forest-covered areas; not only in stagnant water of the developing lake, but also in running water, when the open and wide parts of the Suriname River are compared with the forest-covered narrow streams such as Sara Kreek and Grankreek.

Oxygen values were also different. In the running water of the Suriname River, oxygen was higher than in the Sara Kreek but this was not caused by a difference in temperature or by shadow of trees. Lack of light, turbidity, reductive organic matter and the poverty of green plankton worked together to keep the oxygen content low. Perhaps the most important factor in the Sara Kreek might be the lack of aeration or turbulence, for there were no falls or rapids.

In the stagnant water of the developing lake, oxygen content was influenced by wind action, assimilating green filamentous algae, phytoplankton and organic matter.

In order to get an impression of the variations in oxygen content in open water and forest-covered water, a few samples were taken on April 7, from a cross-section of the lake near Afobaka, from the mouth of the former Sara Kreek and also further upstream. (See also Chapter 4a)

The water had reached a level which covered many small trees and illumination had increased. Oxygen values increased simultaneously and more green filamentous algae were observed between the trees (Table 11).

The observations showed that there was a gradient from left to

90

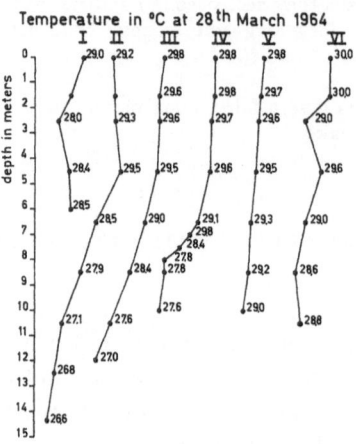

Temperature in °C at 28th March 1964

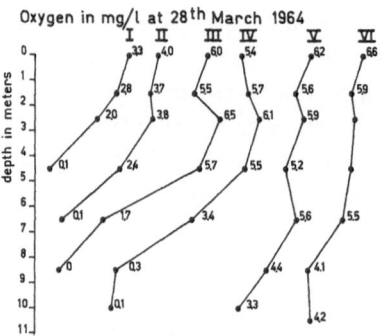

Oxygen in mg/l at 28th March 1964

I = AFOBAKA
II = KOFFIEKAMP
III = ALOESOEBANJA -N   Distance between
IV = ALOESOEBANJA -S   I-II, etc. is ± 25 km each
V = KADJOE
VI = KABEL
4/3 = 4 March, etc.

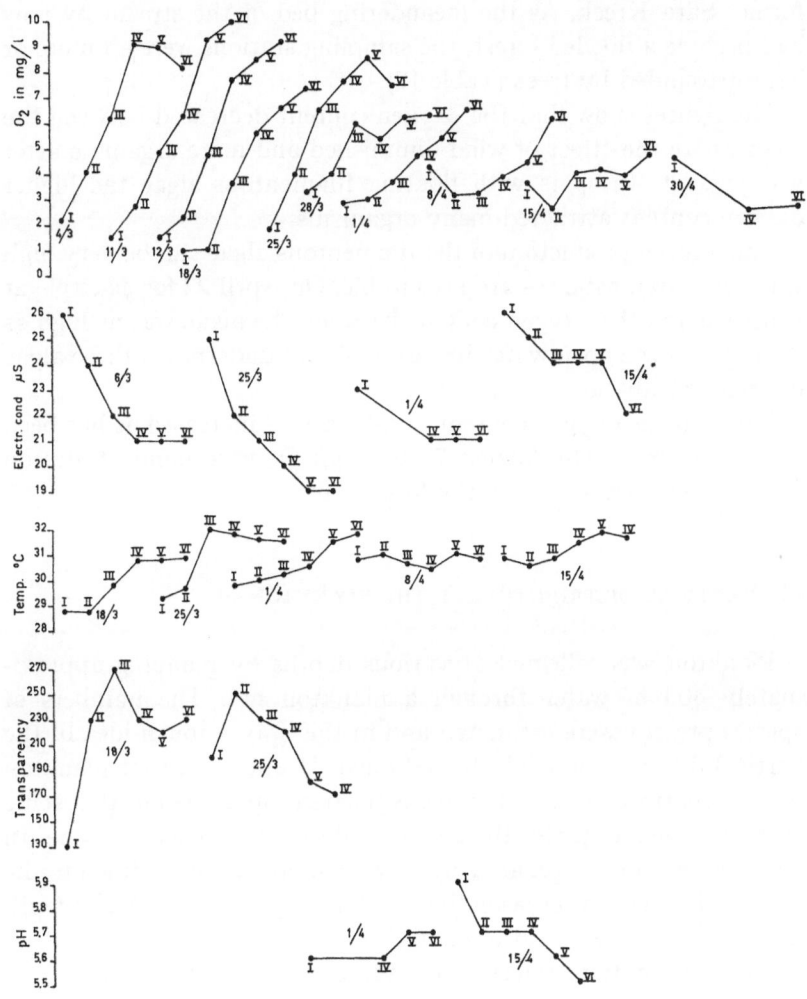

right. On the right side the action of wind was stronger and more green plankton organisms had accumulated there.

During the same period observations were carried out in the former Sara Kreek. As the meandering bed of the stream by now had become a flooded forest, the sampling stations were all more or less surrounded by trees (Table 12).

The figures show that the oxygen content decreased between the trees where the effect of wind diminished and more organic matter was present. In spots with floating filamentous algae the higher oxygen content attracted many organisms.

The oxygen production of the filamentous algae can be very high when the circumstances are favourable. On April 22 for example at Aloesoebanja the oxygen content between the algae was as high as 9.6 mg/l; in the open water just around and underneath these algae oxygen was absent.

Later on the oxygen content at all stations increased as has been described above. On August 27 8.0 mg/l O_2 was found at station Sara at the surface, and in the forest 5.8.

4f. Vertical distribution of the plankton

Plankton was collected at various depths by pumping approximately 40 l of water through a plankton net. The numbers of species present were estimated and in this way a rough idea of the vertical distribution could be obtained. It appeared that plankton was concentrated mainly in the oxygenated top layers but that some species seemed to prefer the layers bordering the anaerobic zone. In a dry period the oxygenated layers were more or less confined to the surface. During a wet season the anaerobic zone sank to greater depth and the plankton was also found deeper.

Some annotations referring to A f o b a k a are given below.

March 26 – Anaerobic zone at 2.5 m. *Eunotia* and *Eudorina* numerous at 1 m. *Conochiloides, Sinantherina* and *Ceriodaphnia* prefer 2.5 m. At 4.5 m only a few specimens found.
April 1 – Anaerobic zone at 2.5 m. Same situation as before. *Eunotia* and *Eudorina* less numerous. A few *Eudorina* at 6.5 m. *Conochiloides* absent in all layers.

May 8 – Anaerobic zone at 2.5 m. *Trachelomonas* and *Strombomonas* dominant and largest numbers at 2 m. *Asplanchna* prefers 1 and 2 m.

June 3 – Anaerobic zone at 6.5 m. Rainy period. *Eunotia, Eudorina, Sinantherina* and others reappear; *Trachelomonas* and *Strombomonas* reduced in numbers. *Trachelomonas, Strombomonas, Conchiloides* and *Moina* only at the surface. *Eudorina* in decreasing numbers to a depth of 5 m. *Eunotia, Conchiloides, Cyclops* and others occur to a depth of 5 m; *Mougeotia* and *Asplancha* increased in deeper layers, most numerous at 5 m.

June 4 – Anaerobic zone at 6.5 m. Same plankton but some species to a depth of 9 to 10 m, such as *Eunotia, Eudorina, Strombomonas, Mougeotia, Asplanchna, Diaphanosoma* and *Moina. Conochiloides, Cyclops, Sinantherina* and *Pedalion* only in the upper 4 m. All species numerous in the upper 4 m; in deeper layers apparently limited by light intensity. *Mougeotia, Asplanchna, Diaphanosoma* and *Moina* present in larger numbers than day before. Of *Trachelomonas* and *Strombomonas* only a few individuals are found.

June 8 – Anaerobic zone at 3.5 m. Same plankton as on June 4, but below 3.5 m only a few specimens.

June 12 – Anaerobic zone at 4.5 m. Rotifers and crustaceans more numerous above the anaerobic zone. *Eudorina* and *Eunotia* decreasing. *Moina* only at 4 m. *Trachelomonas* and *Conochiloides* absent.

June 15 – Anaerobic zone at 2.5 m. Plankton only above 3 m and similar to the plankton of June 8 and 12. No more *Eunotia* and less *Mougeotia*. Total numbers of rotifers and crustaceans lower than on June 12. *Pedalion* only at 3 m.

July 16 – Anaerobic zone at 3.5 m. *Melosira, Conochiloides, Ceriodaphnia, Sinantherina, Diaphanosoma* and *Dictyosphaerium* present above 2.5 m. No difference in numbers at different times of the day at various depths. Practically all numbers decreased with increasing depth. *Cyclops* and *Asplanchna* found down to 3.5 m in the morning, 4.5 m in the afternoon and 5 m at noon. The turbellarian worm *Catenula lemnae* was only met in the zone where the oxygen content dropped to zero. [These fast moving worms could be detected only in living samples; during fixation they desintegrated completely and nothing could be found anymore.] Sometimes thousands of these worms were found together with protozoans such as *Spirostomum* and other turbellarians such as *Microstomum*. On July 16 they were absent above 1.5 m but in the morning thousands were found at 3.5 m; from 2.5 and 4.5 m only about ten were recorded. At noon some 20–50 worms were found at 2.5, 3.5 and 4.5 m but above and below there were none at all. In the afternoon the picture was similar but the worms were more numerous. On July 17 they were practically absent. The observations illustrate the marked preference of these fast moving worms for the borderline between aerobic and anaerobic water.

July 20 – Anaerobic zone at 3.5 m. Roughly the same picture as on July 16. *Melosira* fairly numerous in the morning but only at the surface and at 1.5 m. At noon maxima at the surface and at 2.5 m. *Conochiloides* with maximum at 2.5 m in the morning; at noon mainly concentrated near surface. *Catenula* absent above 1.5 m. In the morning fairly small numbers at 2.5 m. At noon no specimens above 2.5 m; at 3.5 and 4.5 m, however, present in considerable numbers, but absent 6.5 m and lower down.

July 21 – Anaerobic zone at 2.5 m. Very similar to that on previous day. The lowest limit of distribution is now about 1 m higher. Numbers of *Catenula* comparable to those previously found. In the morning only at 1.5 m, at noon only at 3.5 m and in the afternoon at 3.5 and 4.5 m.

July 22 – Anaerobic zone at 2.5 m. All organisms reduced in numbers, the lowest limit of distribution at 1.5 m. *Asplanchna* in the morning and in the afternoon to depth of 4.5 m. An "explosion" of *Catenula* in the morning and afternoon. In the morning from 2.5 to 6.5 m and in the afternoon from 1.5 to 6.5 m. Absent above 1.5 m.

July 23 – Anaerobic zone at 2.5 m. A fair amount of *Cyclops* and *Conochiloides* at the surface and at 1.5 m. A few *Catenula* found below 1.5 m. Other species as numerous as on July 21. The observations in July at Afobaka show diurnal vertical migration of the plankton. Also great variation in numbers is found from day to day to such a degree that it is difficult to account for the disappearance and reappearance of numerous individuals of a certain species, as for instance *Catenula* and some Rotifera. It is quite possible that dense concentrations move in clouds from spot to spot. Nor should we forget that some organisms will die and disintegrate rapidly under the influence of the great diurnal variations in oxygen content. In Chapter 4a an example is given of the disintegration of filamentous algae and the resulting increase of conductivity in a vertical direction. The same may be found for *Catenula* and other organisms which disintegrate readily.

At Kabel the vertical distribution of plankton was examined on a couple of days in May and June. The total population was less numerous than at Afobaka. If we compare the behaviour of some of the species also found at Afobaka, it appears that they may also be found some meters down into the anaerobic zone. It is not easy to see why they should behave like that. In June when oxygen was found fairly deep, plankton was found deep too, with concentrations near the surface and at a depth of 6 to 7 m. Obviously the organisms occurring at 6 to 7 m were more or less trapped in these layers. They were unable to reach the surface in time when the oxygen content diminished and as a result they died. It remains doubtful, however, if this phenomenon is an important part of the mechanism of vertical distribution. Compare Chapter 6.

Finally it should be mentioned that at very sheltered stations such as Sara and Locus, oxygen is totally absent from top to bottom during a long period. Plankton organisms are then only found in a surface layer of a couple of centimeters. As a result net haules contained fewer organisms than samples taken superficially with a bucket. During this period the water is largely covered with a neuston film of iron bacteria and other organisms.

4g. BIOCHEMICAL OXYGEN DEMAND (BOD) (Tables 13–20)

The oxygen content of water depends not only on the physical and chemical properties but also on the activities of plants and animals living in it. Oxygen content is the outcome of a dynamic system of oxygen consumption and oxygen production. If a large amount of organic material or other reductive substances are present in the water, oxygen consumption will be high. Many Bacteria and Protozoa will also develop under these conditions adding to the oxygen consumption.

Examination of water polluted by sewage includes a test of potential consumptive capacity. This test consists of measuring the decrease of oxygen in water kept in bottles in the dark during a certain number of days, and the result is called Biochemical Oxygen Demand (BOD).

During the filling of the Brokopondo lake this test was used to estimate the amount of organic material with easily oxydizable components from the drowning forest. The total amount of organic material has been tested several times with potassium permanganate (Chapter 4h). The values found were low in the river but increased after the impoundment.

Oxygen production of water samples had also been investigated. This has been done by putting BOD-bottles near the window in the room at room temperature and by measuring the oxygen production of the green organisms.

The tests were carried out on several days and at various stations. The BOD at Afobaka was measured in the river proper and later on during consecutive phases after closing up. Several swamps were examined for comparison and also the oxidation pond of the Afobaka settlement built for the purification of waste water.

BOD in the Suriname River (Tables 13–15)

In the Suriname River the oxygen consumption was low. In the bottles kept in the light only a little more oxygen was produced than in bottles kept in the dark (Table 13). In the dark as well as in the light the oxygen content after three days was lower than the initial level. The plankton community of the Suriname River was poor and a limited production of oxygen should be considered normal. In oligotrophic waters in moderate climates oxygen production and consumption in the dark and in the light is also very low (LEENTVAAR 1963). If we take the difference between tests in dark and light as an indication of the capacity for biological selfpurification of the water the conclusion is that this was not very great in the Suriname River, as production in the light was only slightly higher than the consumption in the dark.

In water from the shallow swamps the decrease in the dark was slightly higher than in the Suriname River and the production was in the same order of magnitude. The initial values were different, however. In both cases plankton was also scarce but in the swamps more organic matter was present, reducing the oxygen content. Difference in initial oxygen content between the two swamps of Table 13 is explained by a difference in illumination, and the plankton present.

The saprotrophic oxydation pond was rich in organic matter and also rich in unicellular flagellates such as *Phacus, Trachelomonas* and *Euglena*. The water was turbid. The initial oxygen content was high (even oversaturated) but already after three days the samples smelled of H_2S and oxygen was exhausted. The decrease was more than 9.1 mg/l. In the bottle kept in light, consumption was counteracted by production. The purification capacity of the oxydation pond was insufficient as it was overcharged with organic matter.

On January 9, 1964 another series of samples was taken but they were only kept in the dark (Table 14). The initial values were different from those on November 26; the condition of the water had changed in the meantime.

The series of January 9 showed that the oxygen consumption in the Suriname River was low again after only four days. In the Sara Kreek and the two swamps oxygen consumption was higher. On January 15 a vertical series of samples was taken in the river at Afobaka (Table 15).

The initial (= actual) values of oxygen increased with the depth, as mentioned in Chapter 4a. The decrease of oxygen after three days in the dark was slight in all samples. Near the bottom it was slightly higher, which was to be expected as the silt content was also higher. The figures, however, show that the Suriname River carries very clean water.

BOD in the forming lake (Tables 16–20)

After the closing of the dam large amounts of organic matter started to accumulate. Oxygen consumption increased; the actual oxygen content at the sampling stations decreased. This was observed in the first weeks after stagnation. As the initial oxygen content was very low or even absent, it was not possible to carry out BOD tests in this period. Diluting the BOD sample, a common performance in water pollution research, was not carried out; this would have produced a complete disturbance of the original environmental conditions. Tests in the light were not carried out either.

The decrease in oxygen during this transition period was also very low even in deeper layers. The conclusion, however, that the BOD did not increase after stagnation must be considered to be highly doubtful. Tests in the light were not carried out but it should be mentioned that many *Eudorina*, able to produce oxygen, were present during this period. The actual oxygen content in the lake, however, was very low. The growing amount of phytoplankton, in the beginning, produced an insufficient amount of oxygen to counteract the reductive influences.

The BOD tests were continued in the stabilized periods in May and August, during which the actual oxygen content was higher and restricted to the surface layers. In this period many unicellar green organisms developed. The result showed that the BOD was higher, but oxygen production was not always significant. The figures are comparable to those from the swamps.

In order to understand the underlying processes well, it should be pointed out that the diurnal fluctuations of the oxygen are considerable. The actual oxygen content depends on the time of the day at which the sample is taken and is furthermore influenced by weather conditions.

The rate of oxygen production when samples are exposed to light is shown in the series of May 24, in Table 17. This series also showed that if the test is continued long enough, oxygen content increases after an initial decrease. This was often observed in these tests and it indicates that oxygen production by algae only started after a couple of days. Another explanation of the higher actual oxygen content was the stratification of the top layers which kept reductive substances from moving upwards from the hypolimnion. Furthermore the biocommunity as a whole had gained stability with the result that more oxygen was produced.

The zones of stagnation, transition and restabilization at Afobaka could be followed by sampling in the direction of Kabel where the river entered the expanding lake. With this in mind several BOD tests were carried out during the different phases of the filling up of the lake (Table 18). The results showed an increase of initial actual oxygen content towards Kabel. At Kabel oversaturation occurred. On March 18 heavy rains lowered the values at all stations. Apart from two observations on March 12 the decrease in oxygen content was very small and comparable to that in the river.

This was the expected picture. A complete range was sampled from oversaturated water to water nearly deprived of oxygen. The decrease of oxygen in bottles, however, remained practically the same.

At Kabel samples for BOD tests were taken at several depths every week from February 19 until April 8 (Table 19). This period covered the transition from running to stagnant water. The current ceased on February 19, while after some weeks of supersaturation the oxygen began to decrease after March 18.

The decrease of oxygen in the bottles appeared to be high but in several bottles the initial oxygen content was very high by oversaturation. The decrease therefore is partly due to physical equilibration. In these bottles a bubble of gas was formed under the stopper. In the April samples a lower initial oxygen content was found and oxygen decrease was small.

After restabilization of the oxygen content at the surface at Kabel the decrease in BOD bottles was examined again (Table 20). The decrease of oxygen was comparable to that at Afobaka in May and August in swamps.

The general conclusion is that no pertinent observations on the rate of BOD could be made; also the values found were much lower than expected.

4h. VERTICAL DISTRIBUTION OF ELECTROLYTES AND OTHER CHEMICAL
    DATA (Fig. 29; Tables 21–22)

In the upper reaches of the Suriname River the amount of Cl was
no more than a few mg/l.

As the pH of both river and lake was low, data on alkalinity are of
more importance than the salinity. In the river the alkalinity was
very low (generally about 0.25 mg/l), but a slightly higher value was
found (0.35) near the bottom. The buffering capacity of the water,
therefore, was small. As we have seen in the Suriname River at
Pokigron, the pH dropped during a rainy period. This also happened
with the pH of the lake after stagnation. The equilibrium between
carbonate, bicarbonate and free CO_2 was then disturbed by addition-
al CO_2. In the stagnant lake and in the swollen river the content of
decaying organic material producing CO_2 was high.

It appeared that stagnation alone made the water of the Suriname
River still more aggressive. During stagnation the exchange between
water and air diminished, especially when the water was getting
deep. This might result in accumulation of CO_2.

BRAUN (1952) also found low alkalinities in lakes in the Amazon
region but there the pH increased in a rainy period, possibly because
rain washed away salts from flooded banks.

In May, during the dry season, the alkalinity was higher and it
gradually increased from the surface to the bottom. The concentration of minerals increased likewise. In July, during the rainy season,
the alkalinity decreased. At Sara and Locus both curves ran parallel.
The conductivity mainly represented the amount of carbonates. The
pH was uniform from top to bottom.

The change in colour of the water, after the stagnation, to light
brown by dissolved humic substances suggests the presence of humic
acids. In July a layer of 6 m above the bottom was coloured very
deeply. Since the pH did not change much it was probably not the
amount, but the nature of the dissolved humic substances which was
decisive for the pH level. The level reached soon after stagnation did
not change afterwards despite a constant increase of humic substances. Diurnal changes in pH occurred, however.

Oxydation with $KMnO_4$ suggested only small amounts of organic

VERTICAL DISTRIBUTION OF CHEMICAL AND PHYSICAL DATA IN THE BROKOPONDO LAKE (1964).

29

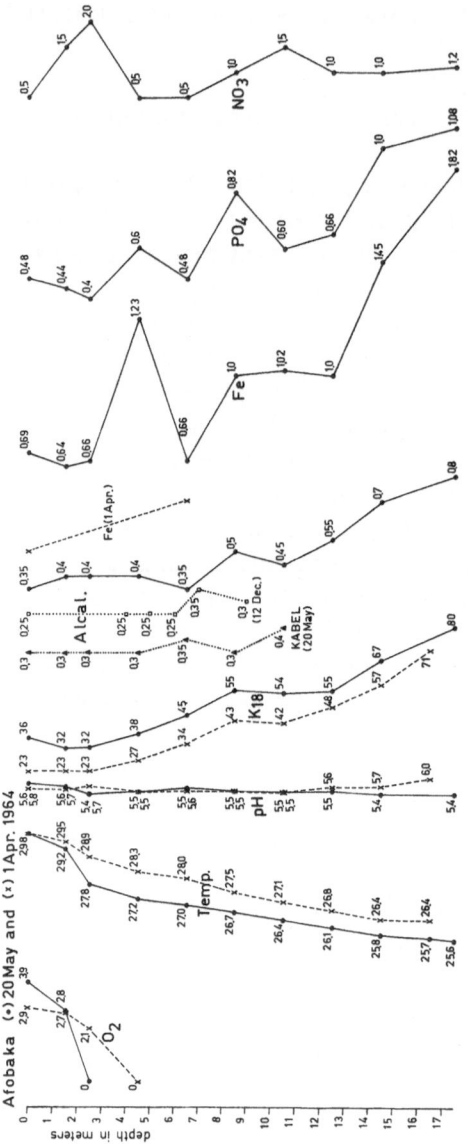
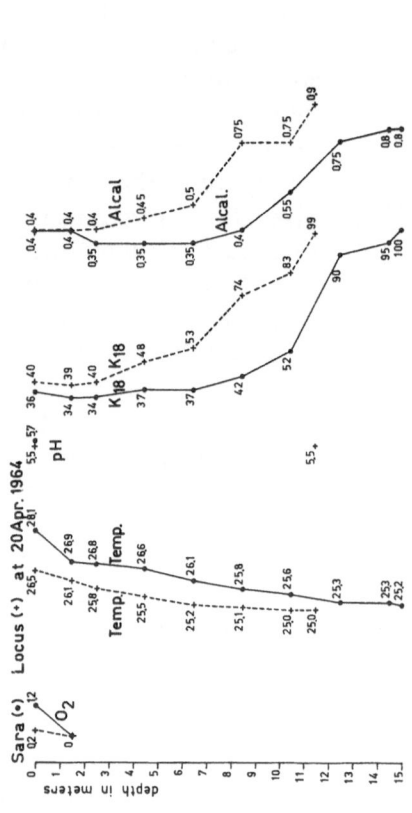

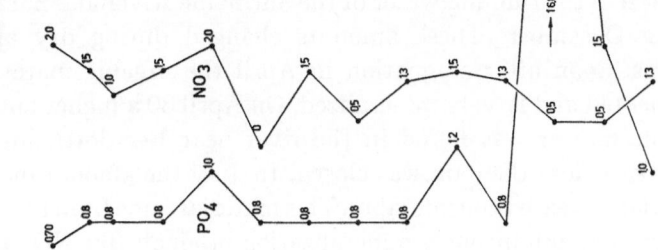

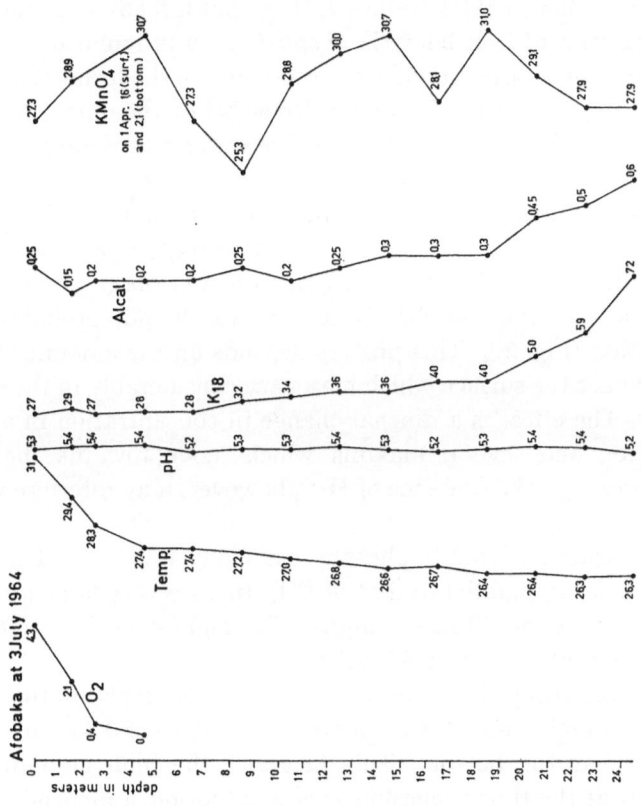

material in the running water of the Suriname River and Sara Kreek during December. These amounts changed during dry and wet seasons. Soon after stagnation in April the organic material was sedimented and largely mineralized. On April 30 a higher amount of organic matter was found in the river near Beerdotti, just as at Afobaka before the dam was closed. In July the amount of organic material increased considerably. This material came from the swamped forest. Simultaneously mineralization products like NO_3 and PO_4 increased too.

A near total absence of N, P and S in the running river was followed by high values after stagnation. More Fe was present than in the Amazon (BRAUN 1952, SIOLI 1964). As has already been mentioned, a thick film of iron bacteria (*Leptothrix*) was found on the water-surface shortly after stagnation; especially in the former Sara Kreek. In the dry bed of the Suriname River below the dam a brown precipitate of iron could be found where water was welling up in little springs.

The iron content increased in May, especially in deeper water layers. In the layer above 2.5 m where oxygen was still present, precipitation of flocky iron hydroxide was observed. Under this layer a maximum of dissolved iron was found, probably due to reduction (Fig. 29). This process depends on the amount of oxygen found near the surface which may vary considerably in the course of a day. The effect is a diurnal change in concentration in a vertical direction and several maxima which move towards the bottom (Chapter 4a). The presence of H_2S, however, may interfere with this process.

The relation of iron to phosphate is shown in Fig. 29. The curve of Fe runs fairly parallel to that of PO_4. In July very large amounts of PO_4 were found (0.70–1.5 mg/l). The highest content (16.8 mg/l) was found at 20.5 m at Afobaka.

SO_4 was always found in the river and later on also in the stagnant water. A high amount was found at Afobaka at 20.5 m in anaerobic water near the bottom. This fact and the high amount of PO_4 present at the time suggests a special situation. Independent movements of the deeper water layers carrying large amounts of sulphate and phosphate will have to be taken into account.

NO_3 also increased after stagnation and reduction below 2.5 m was to be expected. The curve for July gives some indication of a decrease towards the bottom which is just the reverse of what has been found for Fe and PO_4. The curves for NO_3 and $KMnO_4$ are roughly parallel.

Analyses for other electrolytes, such as NO_2 and NH_4, have only been made on samples of running water in December, and of stagnant water in April. Both were absent, or present in quantities undetectable by our methods of analysis.

I shall not try to give a picture of the chemical processes at work in the lake. The data give no more than an impression. At this stage of filling it was evident that the lake water had not yet reached chemical stability and before this happened disturbances would interfere making a reliable picture of interrelations virtually impossible.

The bottom in the upper reaches of the Suriname River was composed of rocks and sand. Soft mud was absent, apart from a few places near the outflow of rivulets and in places where the water was more or less stagnant. In such places many tree leaves could be found also.

The most common type of aquatic macrophytes found were Podostemaceae, occurring in the rapids, *Mourera fluviatilis* being the most common species. In one case a few specimens of *Elodea* were found in a stagnant pool near Djoemoe during the dry season. The river was not suitable as an environment for floating and submerged plants due to the fluctuations in waterlevel and the velocity of the current. For this reason *Eichhornia crassipes* was only found in sheltered places.

The bottom fauna of the river was also very poor. Bottom samples taken with a grab contained only sand. During the dry season, however, several small piles of excrements produced by a species of Oligochaeta could be found along the edges of the sandflats. At this time frogspawn and tadpoles could also be found, deposited by tree frogs in burrows containing water. In the stagnant parts of the river with a muddy bottom and decaying leaves only midge larvae were found. Most organisms seemed to concentrate in the flowing parts of the river especially in the rapids and the falls. In between the Podostemaceae several organisms adapted to strong currents were found. Simulidae and Ephemeroptera were common; Trichoptera, Odonata, Sialoidea (*Corydalis*) and Lepidoptera (*Cataclysta*) were found also.

Freshwater shrimps (*Macrobrachium*) and freshwater crabs (*Potamo-carcinus latifrons*) lived in fissures and quiet parts of the rapids. Small Coleoptera were found mining the emerging leaves of podostemons. Encrusting freshwater sponges were found also. On the rocks snails of the genera *Pomacea* and *Doryssa* were present; their tops were invariably eroded by the aggressive water. These snails were found together with small fish, tadpoles, waterbugs and waterstriders in small pools on the rocks during the dry season. Apart from the few organisms living in stagnant pools all specimens belonged to groups normally living in running water.

The presence of a veliger-type of larvae and zoecia in plankton samples indicated that Lamellibranchiata and Bryozoa must be present in the river. Adult colonies of Bryozoa were not found, however. Only a few shells of *Diplodon voltzi*, *Diplodon granosus*, and *Castalia ambigua* were found occasionally. The veliger larvae in the plankton probably belonged to *Diplodon voltzi*, the most common lamellibranch in the Suriname River. Shortly after the dam at Afobaka was closed a large number of *Diplodon voltzi* could be collected from the dry bottom of the lower course. The mussels were present especially in the soft muddy parts near the outflow of small tributaries. Freshwater shrimps were concentrated in the shallow parts of the dry riverbed filled with stagnant water; their presence was betrayed by numerous little pits in the sandy bottom. Worms could be collected easily in places where excrements were found, piled on the surface.

The observations in the dry riverbed gave a certain indication of the food chain of fish in the river. The scarcity of food organisms was obvious. Probably only the freshwater shrimps were of some importance as a source of food as they were most numerous and formed an easy prey. The majority of the fish were predators. Insects blown into the water from the banks may have formed another source of food, though I never observed fish feeding on floating insects.

After the closing of the dam the water became stagnant and all bottom organisms must have died rapidly through lack of oxygen. This has already been mentioned in connection with the catfish, *Plecostomus plecostomus*, dependent as it was on the rapids. Other bottom-dwelling organisms which could not escape to more favourable places also died. Some sponges, freshwater shrimps, ephemerids and dragonflies could be found on floating substances or other places where oxygen was present. Several organisms, common in stagnant water, also invaded this substrate. This is apparent from the fact that in the Sara Kreek region extensive fields of *Lemna valdiviana, Spirodela biperforata* and *Ceratopteris pteridioides* developed, which must have originated in stagnant water. In the flooded forest dense algal mats (*Spirogyra, Mougeotia*) developed, together with many organisms from stagnant water like the snails *Drepanotrema anatinum, Aplexa marmorata, Acroloxus, Gundlachia, Taphius kühnianus*; the crustaceans *Euryalona occidentalis, Chydorus, Cyclestheria hislopi*; ephemerids, Odonata and waterbugs. In the Sara Kreek and other shallow rivulets with mainly stagnant water *Ranatra linearis* and other Rhynchota could be found (*Belostoma, Microvelia*). The snails mentioned were only present at certain localities from where they were dispersed with the floating material. During the rainy season this bio-community did not survive, many dead *Pomacea glauca* were found floating between the duckweed.

The dead leaves of the swamped trees were soon covered with green masses of filamentous algae (*Spirogyra, Mougeotia*), which eventually died when the water level rose. On the dead leaves of the trees many muddy tubes of Oligochaeta were visible (*Dero, Aulophorus, Aeolosoma* and *Pristina*).

Where filamentous algae developed, bubbles of gas kept the mats of algae floating. In places where dense mats of Lemnaceae developed the algae died from lack of oxygen and light. In the former Sara Kreek region where oxygen was absent from top to bottom, the floating algae formed oxygen-producing islands favourable for many organisms. Crustaceans and young fish were seen just around these patches of algae.

Among the roots of waterhyacinths a rich community of organisms could be found. It was composed mainly of crustaceans, filamentous algae, desmids, midge larvae, Odonata, ephemerids, shrimps, waterbugs and small fish. Also larger fish were found, such as *Gymnotus*, hiding among the roots from where they could easily reach the surface.

COMPARISON OF THE FAUNA AND FLORA OF THE SURINAME RIVER
before and after the closing of the dam (Nov. 1963 ↔ Sep. 1964)

SURINAME RIVER (running water)	BROKOPONDO LAKE (stagnant water)
Plankton mainly composed of phytoplankton	Plankton of another type: other phytoplankters dominant; Cladocera and Copepoda in great numbers in oxygenated top layers; new species from stagnant waters (e.g. *Cyclestheria, Euryalona; Lacinularia, Octotrocha, Sinantherina*).
Crabs and shrimps (*Macrobrachium*), and other bottom-dwelling organisms.	Crabs and shrimps, and other bottom-dwelling organisms only near the oxygenated surface.
Velidae, Gerridae, and Belostomidae scarce.	Velidae (*Microvelia*) in great numbers; Gerridae not observed; Belostomidae on floating material only.
Trichoptera common on rocks in the rapids.	Trichoptera not recorded.
Ephemeroptera common on firm substrates.	Ephemeroptera not common; only a few species (*Asthenopus, Callibaetes*) recorded.
Megaloptera, Lepidoptera (*Cataclysta*), and Simulidae.	Megaloptera, Lepidoptera, and Simulidae not recorded.

Sponges on rocks.

Sponges on floating substrates only.

River molluscs (*Pomacea, Doryssa; Castalia, Diplodon*).

River molluscs absent; species from stagnant water locally and temporarely (*Acroloxus, Aplexa, Drepanotrema, Gundlachia, Taphius*).

Earthworms common; no Naidae recorded.

Earthworms not recorded; *Dero* and *Aulophorus* on decaying substrates.

Turbellarians not recorded.

Turbellarians numerous in the plankton (*Catenula, Mesostoma*).

Fishes in many species and specimens.

Fishes in only a few species and specimens; surface dwellers (*Leporinus*) in oxygenated top layers; Gymnotidae among floating plants.

Podostemaceae common in rapids.

Podostemaceae absent.

Eichhornia and *Ceratopteris* in only a few specimens.

Eichhornia and *Ceratopteris* covering large areas.

Duckweeds not recorded.

Duckweeds (*Lemna, Spirodela*) locally covering large areas.

Filamentous algae scarce.

Filamentous algae locally forming mats.

Blue-green algae on firm substrates.

Blue-green algae abundant on roots of *Eichhornia* and on other floating substrates.

6.   APPENDICES

OBSERVATIONS IN 1968

Hydrobiological research of the Brokopondo Lake started (some time before the dam was closed on February 1, 1964) with weekly observations at various stations in the vicinity of the dam. When the last members of the research team left around the middle of 1967, these periodical observations were stopped, although the water of the Lake had not yet reached its highest level and a stable biological situation had not yet been established.

The Division of Hydraulics (WLA) of the Ministry of Public Works and Traffic at Paramaribo continued monthly observations at five stations in the Lake, including observations on the rainfall. They collected information on the pH, conductivity, turbidity, oxygen content and the temperature at depths of 0, 1.5, 2.5, 3.5, 4.5, 6.5 and 8.5 m, while plankton samples were taken as well. This information and the samples were regularly forwarded to The Netherlands, for study by research-workers of the Brokopondo-team. The kind help of Ir. A. BLES of WLA is gratefully acknowledged.

The author revisited the Brokopondo Lake from April 1 to May 8, 1968; the results of his observations follow below.

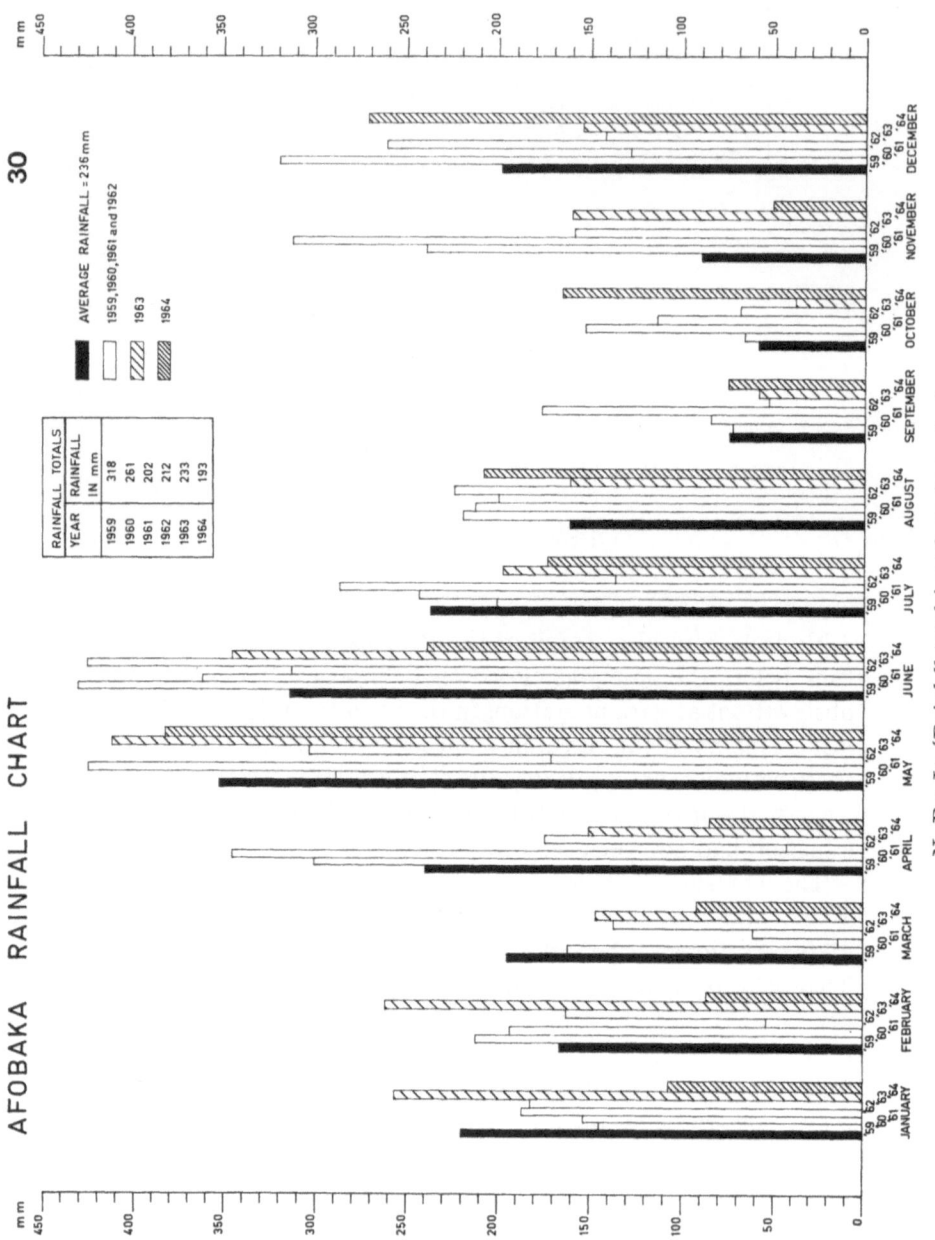

AFOBAKA RAINFALL CHART

30

AVERAGE RAINFALL = 236 mm

1959,1960,1961 and 1962

1963

1964

RAINFALL TOTALS	
YEAR	RAINFALL IN mm
1959	318
1960	261
1961	202
1962	212
1963	233
1964	193

N. B.: In 'Rainfall totals' read cm instead of mm.

WATERLEVEL

On April 30, 1968, the water gauge at the dam read 237 ft (41.80 m). Since April 1 the water had risen 5 ft (1.50 m). In 1967 the highest level recorded was 241 ft (43 m) above Surinam level.

About 40% of the lake was covered by *Eichhornia*. According to WLA the evaporation of the free water surface at the dam amounted to an average of 4.5 mm per day. The precipitation at Afobaka averaged 5.5 mm and at Pokigron 5.7 mm.

CHLORIDE CONTENT

In the older parts of the Lake a slight increase of the Cl-content had been observed (from 5 to 31 mg/l). Considering that Cl is not influenced by biological processes and that the rain water and the river water contain only a few mg per liter, one wonders what the cause of this increase could be. In a closed basin the Cl-content increases when the evaporation exceeds the precipitation. Apparently more Cl had been supplied by creeks and rivers than was carried off through the turbines.

WIND EFFECT

The increasing effect of the wind on the Lake also increased the mixing of the water layers, influencing the oxygen supply in the water. Consequently the rotting of the organic material proceeded more rapidly than anticipated. The increase of the wind effect was due to the control of the waterhyacinth fields and to the disappearance of tree tops below the water surface. As the wind and the action of the waves swept waterhyacinths away, there was much open water near the dam reaching as far as Kabel, Beerdotti and Locus. Waterhyacinths, however, still occurred between the trees and further south.

The waterfern (*Ceratopteris*) appeared to be absent and could probably not stand the increased light intensity. The waterhyacinth

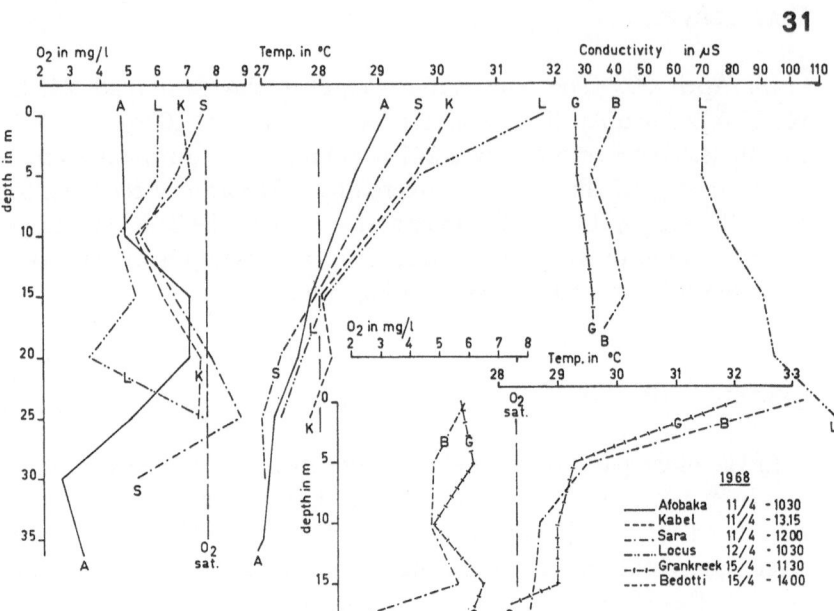

31

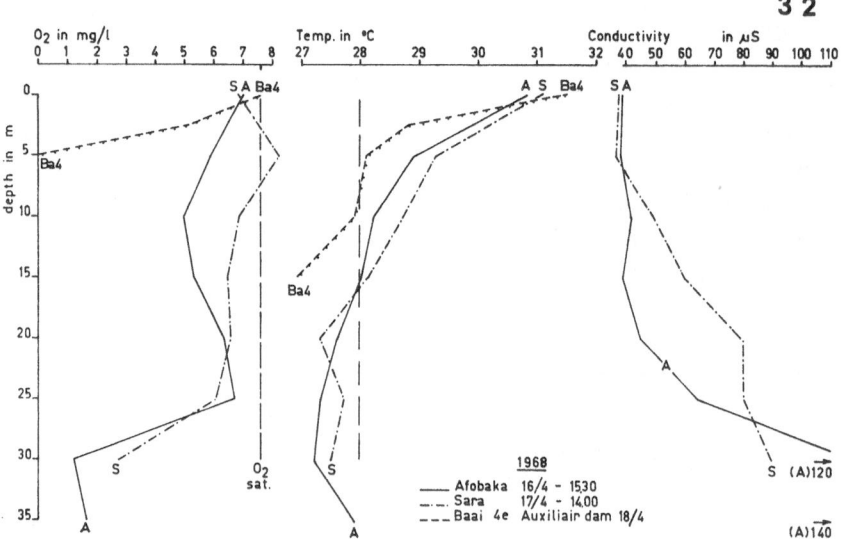

3 2

(*Eichhornia*) has been controlled by spraying with 2–4 D from an aeroplane and from boats, the costs amounting to about a quarter of a million dollars a year (VAN DONSELAAR, 1968). Sport fishing has become quite popular, as toekoenari (*Cichla ocellaris*) and pireng (*Serrasalmus rhombeus*) are plentiful.

FIELD DATA (Figs. 31–36; Tables 23, 26)

The red station markings were clearly visible everywhere, as they had been re-painted periodically by WLA during the rise of the water.

At station Afobaka, on April 9, oxygen was present down to a depth of 24 m and the O_2 content was higher in deep water than nearer the surface. This situation was rather unexpected as in previous years oxygen had only been found in the upper 4 to 5 m layers and occasionally in a low concentration down to 8.5 or 10 m. In former times H_2S was usually present in the deeper water layers. This time no H_2S could be found at Afobaka, but the river below the dam smelled clearly of H_2S and contained no oxygen.

On April 11 and 12 oxygen content and temperature were measured at the stations Afobaka, Kabel, Sara, Locus and the railway track at Kabel, every 5 m in a vertical series down to the bottom. The result is shown in Table 23–24, Fig. 31.

The information gathered on temperature and oxygen showed that there was some oxygen at all stations. The temperature gradually dropped towards the bottom.

At Afobaka the water near the bottom, at a depth of 35 m, contained 3 mg/l O_2, and at Sara there was supersaturation at a depth of 20–25 m. This means that a definite change had taken place in the oxygen content in comparison with the situation of 1967 and previous years. Although this phenomenon is still unexplained, this increase in the oxygen content at all depths and at all stations may indicate that most of the organic matter in the lake water has been mineralized and putrefaction has nearly come to a stop.

The question remained why, in spite of the presence of oxygen in all water layers especially near the dam, almost anaerobic water flowed through the turbines and below the dam the river contained hardly any oxygen (according to observations at Brokopondo, Phedra and Berg-en-Dal).

It was found that water samples taken from the deeper layers in the lake smelled of H_2S in spite of the oxygen present. There were also many small gas bubbles in the samples, formed because of reduced pressure in the water (Table 25). The presence of both H_2S and O_2 in deep water layers has been recorded also from lakes outside Surinam.

The BOD measurements were interesting in this respect (Table 23, Fig. 33). At all stations the oxygen content decreased faster in the layers below 5 m than from zero to 5 m. Clearly putrefaction in the upper layers is less than below 5 m. This is similar to the situation in former years, when a transition from anaerobic water to water containing much oxygen at a depth of about 5 m was recorded, together with a sudden change in temperature. As well as the water samples taken at the

114

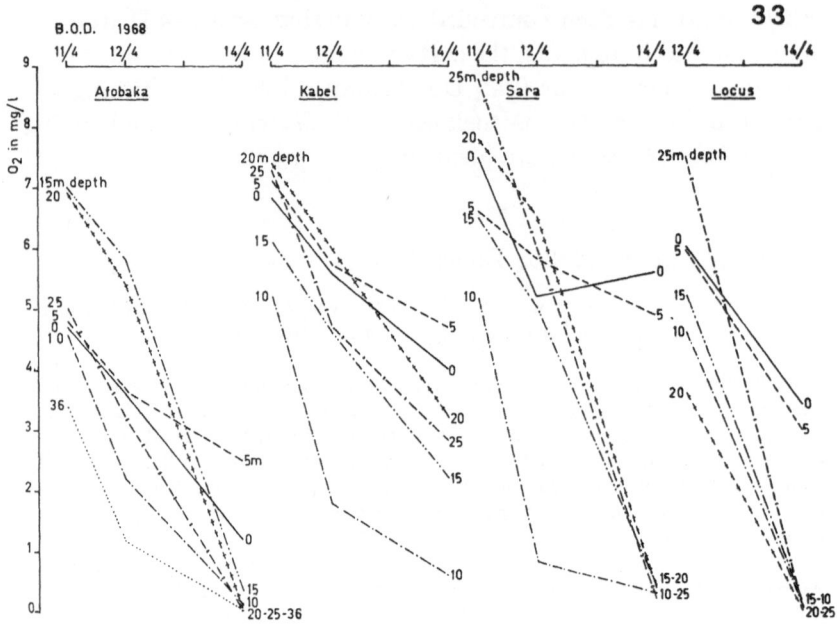

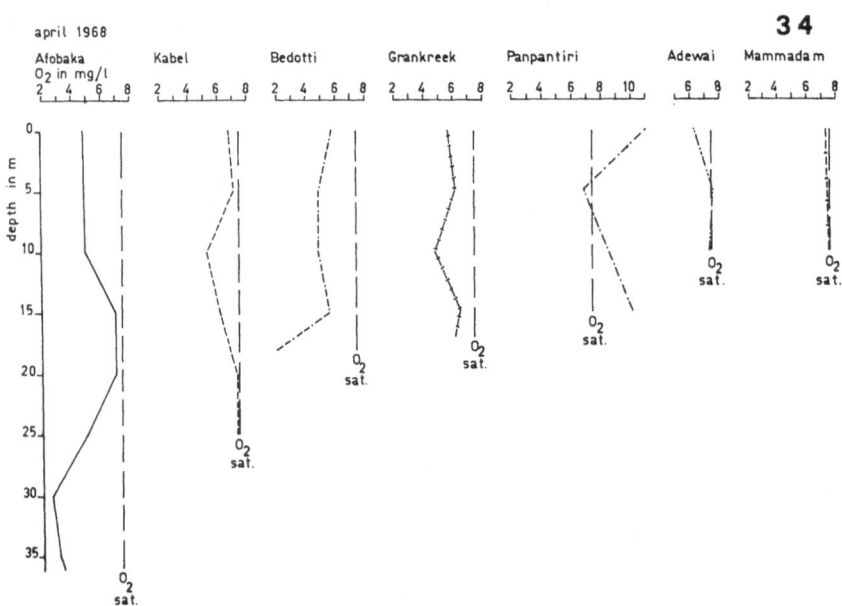

stations at Kabel, Afobaka, Sara and Locus, all of which were situated in open water above the former river bed, samples were taken at shallow, sheltered places such as the former railway track at Kabel and at Sikakampoe, situated about an hour's sailing beyond Locus; waterhyacinths grow at both places, and shelter is offered by many dead trees.

The oxygen content here was lower than anywhere else, but the temperature was higher, with a sharp decrease towards the bottom. Apparently there was little mixing of the water layers owing to the absence of wind effect. At Sikakampoe the oxygen demand was considerable, both at a depth of 5 m and near the surface. At the site of the railway track the water was shallow and near the bottom the oxygen decrease was less than near the surface. Reduction at both stations was greater, probably owing to the growth of waterhyacinths and concentration of organic matter; there was also a stronger growth of filamentous algae.

A pronounced decline of temperature towards the bottom, connected with lack of wind, was also observed on April 18 in the bay near the auxiliary dam No. 4 at Afobaka, where the water below 2.5 m lacked oxygen. On the other hand, there was a less pronounced temperature fall at the places where samples has been taken on April 11 (Fig. 32) and where an increasing wind effect, connected with the gradual disappearance of trees and branches, had mixed the water to a greater depth than in the bay of the auxiliary dam.

At the mouth of the Grankreek in the former Suriname River and at Beerdotti oxygen was also present down to the bottom (Table 23). The plankton in this water did not overproduce oxygen as it does in hypertrophic water. None the less there was a copious development of green phytoplankters, which are capable of producing oxygen.

At Grankreek and Beerdotti more oxygen was present at a depth of 15 m than in the higher layers. At Kabel on April 21 there was a pronounced layer with oxygen at a depth of 15 m only. The same was true at Sara and Afobaka on April 16 and 17. Presumably this was surface water which had absorbed oxygen, pressed downwards by the wind below the surface.

The BOD-light and BOD-dark experiments clearly show the difference in the reductivity condition of the water above and below the 5 m line, especially at Sara and Afobaka. The actual oxygen content of the water does not have much significance if one does not know anything about the reduction and production processes of the water. The oxygen content in the upper 5 m is quite stable, below that depth it is unstable in all layers; under certain circumstances, for instance by mixing after passing through the turbines, it may disappear rapidly. Mixing caused by wind and rain may have the same effect.

It may be expected that gradually an oligotrophic environment will be established, depending on the nature of the bottom of the lake and the character of the water supplied. The increase of desmids which are characteristic of this environment, is an indication of this, and it may be expected that a stable oxygen content will be formed, amounting to a few mg/l near the bottom. Within a few years the river below the dam will probably have more oxygen and more fish and other organisms.

In the transitional zone at the Southern side of the lake, where the Suriname River empties into the lake, samples were taken on April 23, 24 and 25 (Fig. 36, Table 23). At Pokigron the water-gauge was high because of heavy rainfall (4.40 m). The

3 5

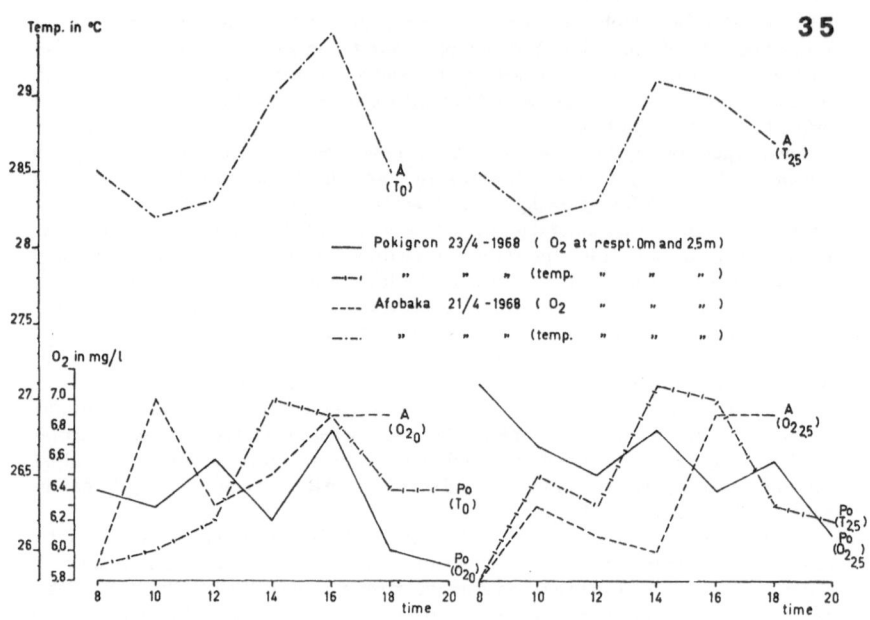

3 6

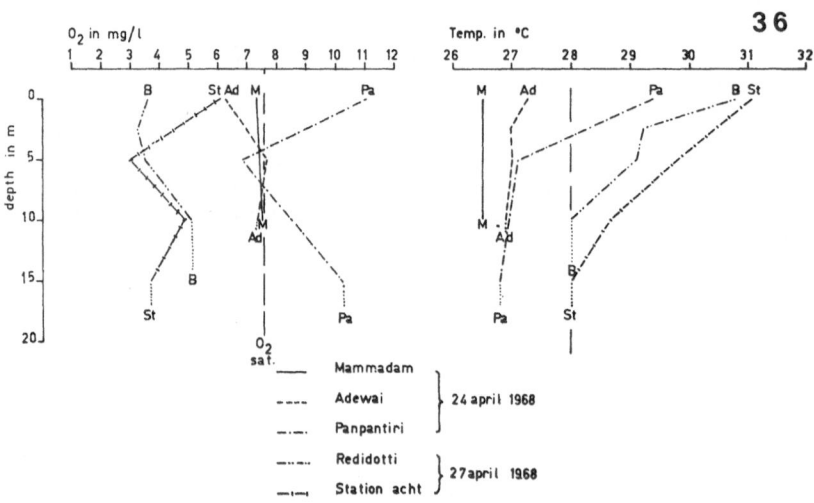

Mamadam was under water and here too dead parts of the forest were found, but the water was not altogether stagnant. Samples were also taken at Adewai and Panpantiri. The river temperature was lower than the temperature of the lake and it rose towards the North. At Mamadam the water was almost saturated with oxygen down to the bottom. In the stagnant water at Adewai the oxygen content was higher in the deep water than nearer to the surface, while it was supersaturated at both the bottom and the surface at Panpantiri. The zone at Panpantiri probably had the strongest biological selfpurification. Changes could also be observed in the composition of the plankton and the colour of the water, which was more brownish here. The water was shallower, the trees were denser and higher, so that the wind effect was slight.

In Fig. 34 (Table 23) the oxygen content at the various stations in a north-south section is shown. In the southern part of the Lake the water below 5 m was not anaerobic. Here too putrefaction has diminished. The rise of the waterlevel has been slight, no new parts of the forest have been inundated and no more trees have died off. The supply of organic matter has diminished. As the size of the lake is very large, at present, the dying of the forest will occur mainly at the shores and the same is true for the supply of organic matter, which may manifest itself locally through the development of filamentous algae or plankton. Apparently only a pronounced rise in the water-level will give an extra supply of organic material which will affect the central and northern parts of the Lake.

At Pokigron the oxygen content and temperature at the surface and at a depth of 2.5 m were measured every two hours in the flowing river (Fig. 35). The location was situated not far from the bank and the bottom was sounded at 3.10 m. The highest temperatures were measured between 14 h and 16 h. The oxygen content at the surface of the river was highest at 16 h. At a depth of 2.5 m it was highest in the morning. In the stagnant water of Afobaka the conditions were irregular.

In the area of the former Sara Kreek samples were taken at station 8 at Redidotti, about 3 hour's sailing from the dam, not far from the former village of Dam. The water was stagnant here and dense fields of waterhyacinths occurred everywhere. Fig. 36 shows that the oxygen content was relatively low in open water; a peak was found at a depth of 10 m.

DAILY FLUCTUATIONS (Fig. 37; Tables 24–25)

Temperature and oxygen content fluctuated during the day. At Afobaka observations were made on April 21 every two hours (Table 25, Fig. 37). In the upper 2 to 3 m there was always much oxygen. In the deeper layers this was not always the case, although one or two layers may have a higher O_2 content than those directly above or below.

It appeared that a relatively warm layer in the deep water also had a relatively high oxygen content. It is curious that a certain

37

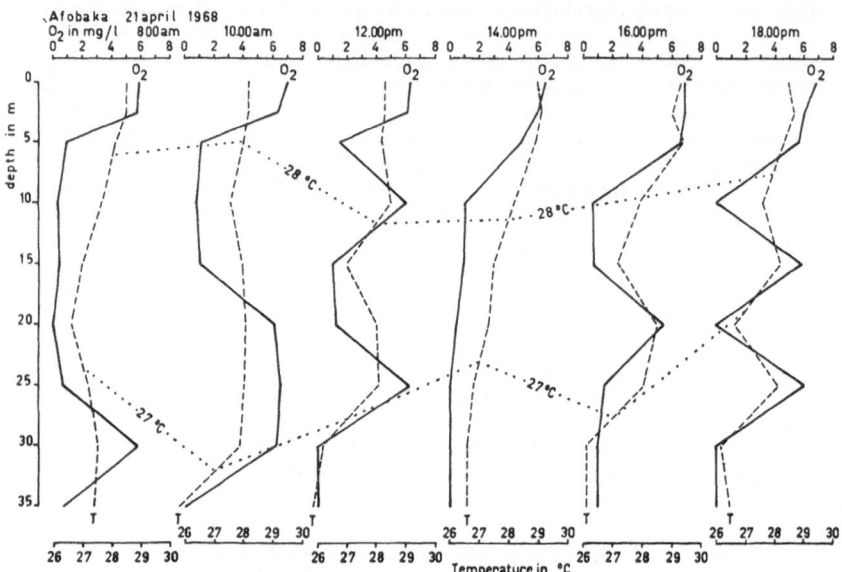

38

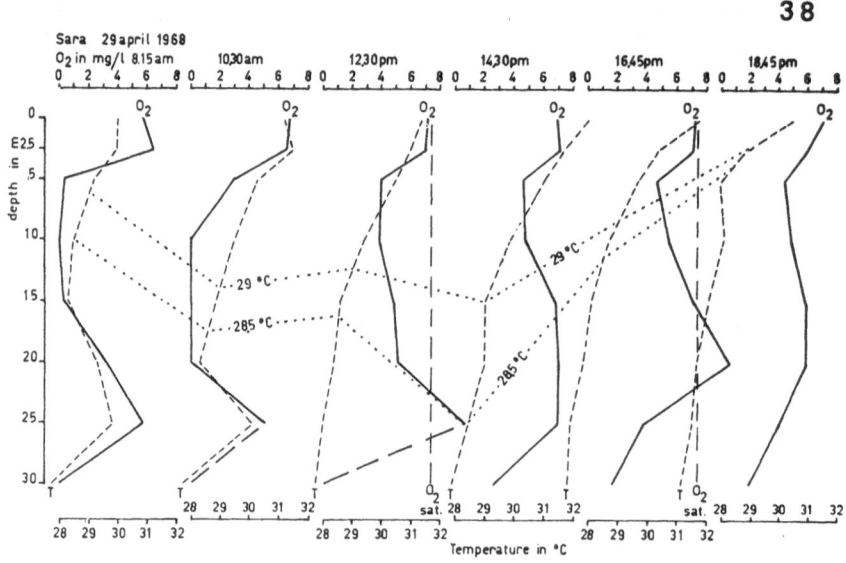

layer in deeper water should be rich in oxygen, since it is known that in the absence of light there is no oxygen production.

Because of the continuous rainfall, heating of the surface layers only occurred at 14 h and 16 h, when the sun was shining. During the rainfall in the morning the wind was strong, causing mixing of water (Fig. 37): at 12 h the 28°C isotherm drops to a depth of 12 m.

On April 29 measurements were taken at station Sara, where there was no influence of currents, such as caused by the turbines of Afobaka. On that day there was constant sunshine and also wind. The water was warmer here; for comparison with Afobaka Fig. 38 shows the 29° line. The heating of the upper water layers began at 10.30 h and penetrated deeper, probably by wind effect. Oxygen supersaturation was present at a depth of 20 m at 16.45 h and at 25 m at 12.30 h, which may be explained by the downward transport of supersaturated surface water by currents; since there is a simultaneous increase in pressure a relatively large amount of oxygen remained in solution.

Between 10.30 and 12.30 h there was a remarkable difference in oxygen content in the layers between 5–20 m; perhaps due to a supply of oxygen-rich water from elsewhere. For example, at 8 h and 10.30 h, water was found with a temperature higher than 29°C at a depth of 20–25 m (see the points of intersection), which was not found at 12.30 h later. The rapid changes in temperature and oxygen in the course of the day might be explained by turbulent currents, caused by increased winds. Because of the oxygen and temperature conditions observed, the monthly program executed by WLA was altered. From April 30 onwards samples were taken every 5 m down to the bottom (Table 26, Fig. 39). The pH at Beerdotti was constantly lower, and often there was a higher pH in the upper 5 m.

The conductivity was lowest at Beerdotti, at all depths. It increased at Kabel and Afobaka. Below 20 m it increased strongly at Afobaka, while at Sara and Locus its increase downwards surpassed that at the other stations, on account of the environment (the stations Sara and Locus were situated in denser inundated forest and the water had a higher conductivity here).

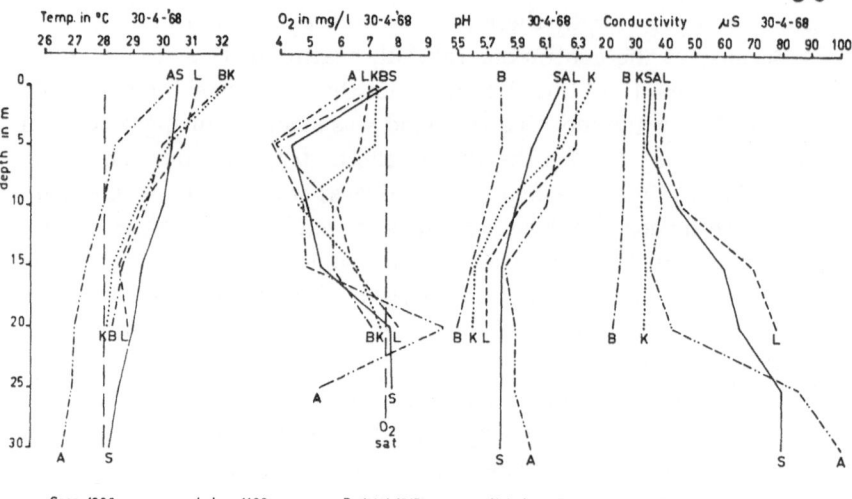

| Temp. in °C 30-4-68 | O₂ in mg/l 30-4-68 | pH 30-4-68 | Conductivity μS 30-4-68 |

———— Sara 10.30am ———— Lokus 11.30am —·—· Bedotti 13.15pm .... Kabel 15.45pm —··—··— Afobaka 16.45pm

During the observations a note was made whether living or dead Crustacea were found, in order to learn whether the plankton was influenced or not by rapid changes in the oxygen content (Tables 24 and 25). Dead animals were found in some samples from water without oxygen. Plankton was found in all water layers but mainly in the upper 5 m of water. At a depth of 10 m there were much fewer plankters. At Sara below a depth of 10 m the specimens were mostly dead or inactive; at a depth of 15 m plankton was sometimes absent. The situation at Afobaka was different; living organisms were found at all depths below 10 m. Probably there was more disturbance in the water at Sara than at Afobaka. At all depths the mosquito larva *Chaoboris* was collected.

PLANKTON (Tables XV–XVII)

Table XVII gives the composition of the plankton from Afobaka to Pokigron, and in the area between the former Sara Kreek to Redidotti. The flowing water of the Suriname River contained few species and specimens. Brown detritus was fairly abundant, with spiculae of sponges, iron bacteria and colonies of the flagellate *Rhipidodendron huxleyi*. The plankton had a brown colour. At Mamadam we found about the same river plankton, although the current had almost ceased. Half an hour's sailing further north at Adewai the water was stagnant, with little plankton, in which most of the species of potamoplankton were lacking; a slight increase in zooplankton was observed. The carplouse (*Argulus*) was found in the plankton. At Panpantiri the colour of the plankton changed to pale green because some species, such as *Volvox*, *Eudorina* and *Cosmarium*, became abundant.

The pronounced supersaturation of oxygen at Panpantiri, mentioned above, indicated a strong biological self-purification. At the mouth of Grankreek such a supersaturation could not be detected but the composition of the plankton differed from that of the lake environment further on. *Cyclops* and *Ceriodaphnia cornuta* were present in great quantities, as were *Cosmarium* and *Eudorina*.

Nearing the dam all these species diminished in numbers and the plankton was mainly composed of large quantities of Desmidiaceae, such as *Staurastrum leptacanthum*, while Crustacea, Rotifera and, to a lesser extent, *Volvox* and *Eudorina* were present. The plankton also contained many Ostracoda. The plankton of the lake environment consisted mainly of desmids and, in minor quantities, Crustacea, Rotatoria, *Volvox* and *Eudorina*. Diatoms and blue-green algae were not found. In the neuston the green alga *Botryococcus braunii* occurred mainly in the surface water.

The strong development of Desmidiaceae indicated that the lake water probably had become less saprobic. The data on the biochemical oxygen demand and the decrease of the flagellatae *Eudorina* and *Volvox* point in the same direction; in previous years both species were present in great numbers. Considering that the Desmidiaceae still developed in huge quantities, one may conclude that as yet there was no balance in the environment. As soon as this balance is reached, it is expected that gradually an oligotrophic environment will occur, with Desmidiaceae dominating, although in somewhat smaller numbers; other species will join them.

On the railway track at Kabel the observation point was not far from the bank. The depth was only 2 m and there was an abundance of waterhyacinths; huge quantities of *Volvox*, *Eudorina*, *Micrasterias* and Crustacea were present. The species of open water such as *Staurastrum leptacanthum* were only present in small numbers. The species of *Micrasterias* and also filamentous algae and Desmidiaceae occurred in shallower water with much waterhyacinth: Station Redidotti in the Sara Kreek and Sikakoempoe, Station 8, Locus and Sara – with the open water area increasing – resembled the lake environment at the dam and Kabel.

On the Western side of the Suriname River at Panpantiri old waterhyacinth fields contained many aquatic insects, such as dragonfly larvae, waterbugs (*Belostoma* and others), waterbeetles; also young fish, the Crustacea *Cyclestheria hislopi* and one or two snails of *Drepanotrema* cf. *anatina*. Of the macrophytes *Utricularia* and *Jussieua natans* are mentioned. In the waterhyacinth vegetation of the railway track at Kabel fewer types were found. The snail *Drepanotrema* was found somewhat more frequently. Dragonfly larvae were abundant, as well as the shrimp *Macrobrachium*. There was also more growth of mats of blue-algae.

We have the impression that the old waterhyacinth fields had a greater variety of macrofauna than the young fields. In the sprayed fields and in those found at Sikakoempoe there appeared to be fewer species.

OBSERVATIONS IN 1972

During a one day's visit to the Lake, May 3, 1972, the reservoir was sampled at 5 stations. These stations are situated at Afobaka, Sara, Kabel, Beerdotti and Locus, while the Department of Hydraulics of the Ministry of Public Works and Traffic (WLA) at Paramaribo carries out monthly samplings. It might therefore be useful to give a view on the present situation in the lake.

The lake was filled up, and since July 1971 the spillway has been used when needed. This has resulted in irregular floodings of the Suriname River downstream the dam, which has given troubles to the people living there. Just below the dam, the river still smelt of H_2S.

As many trees in the lake disappeared under the water, there was a very extensive area of open water, without any growth of floating vegetation. The waterhyacinth appeared to be eradicted by the sprayings of 2–4 D, and managers of the lake hoped to be able to cope with any new invasion of floating vegetation now. On several shallower parts of the lake concentrations of dead trees could be found with some epiphytic plants on it, and many webs of spiders which were the only animal alive above the water. Birds were not observed. The effect of winds had increased and periods of absence of wind were followed by sudden storms during showers, causing high waves.

Temperature and oxygen content observations from May 1972 and the monthly observations of WLA in 1971 show the following: There is often no distinct thermocline, but oxygen is always present in discontinuity layers. The well oxygenated epiliminion extends to a depth of 5 or 10 m, the metaliminion from 5–15 m. This general pattern was found at each of the 5 stations. In comparison with the observations of 1968, there was no improvement of the oxygen content of the deeper waterlayers. This might be due to the effect of the rigorous killing of the waterhyacinths, which sank down to the bottom, prolonging the period of rotting of the lake (LEENTVAAR 1974).

BOD$_5$ was only a few mg/l above 25 m. Chemical analyses from samples taken at the surface and at 25 m in the Lake near the dam, and another from the Suriname River near the outlet of the turbines showed low contents of Ca^{++} (1.5–3.5 mg/l), PO$_4'''$ (0.02–0.07 mg/l), absence of NO$_3'$, but the Fe content was rather high (0.18–0.95 mg/l).

The composition of plankton did not differ much from that in 1968.

124

7.   REFERENCES

ABDIN, G., 1949. Biological productivity of reservoirs. *Hydrobiologia 1*, p. 469–475.
ALLSOP, W. H. L., 1960. The Manatee: ecology and the use for weed control. *Nature 188 (4752)*, p. 762.
ALTEMÜLLER, H. J. & KLINGE, H., 1964. Micromorphologische Untersuchungen über die Entwicklung von Podsolen in Amazonasbecken. In: *Soil micromorphology*, p. 295–305.
BARD, J., 1964. Les travaux de l'Université de Liège sur le Lac Mwadingusha (Katanga). *Revue bois et forêts des Tropiques 96*, p. 68–92.
BERTRAM, C., 1963. *In search of Mermaids. The manatees of Guiana.* London, 183 pp.
BERZINS, B., 1961. New Rotatoria from Victoria, Australia. *Proc. R. Soc. Victoria 74*, p. 83–86.
Biological Brokopondo Research Project, Surinam. *WOSUNA Report for the year 1963*, p. 20–22.
Biological Brokopondo Research Project, Surinam. See: *Progress reports.*
BOESEMAN, H., 1968. The genus Hypostomus Lacépède, 1803, and its Surinam representatives ... *Zool. Verhand. Leiden* 99, 89 pp.
BRAUN, R., 1952. Limnologische Untersuchungen an einigen Seen im Amazonasgebiet. *Schweiz. Ztschr. Hydrol. 14*, p. 1–129.
Brokopondo Research Report. Found. Scient. Res. Surinam and Neth. Antilles, Utrecht. In press.
CARTER, G., 1934. Freshwaters of rain-forest areas of British Guiana. *J. Linn. Soc, London (Zool.) 39*, p. 147–193.
COCHE, A. G., 1968. Description of physico-chemical aspects of Lake Kariba, an impoundment in Zambia-Rhodesia. *Fish. Res. Bull. Zambia 5*, p. 200–267.
DAMAS, H., 1960. La Lac de barrage de la Lufira. *VIIe Réunion Techn. de l'UICN Athènes 4*, p. 152–157.
DEMOULIN, G., 1966. Contribution à l'étude des ephéméroptères du Surinam. *Bull. Inst. R. Sci. Nat. Belg. 42 (37)*, 22 pp.
DONSELAAR, J. VAN, 1964–1969. In: *Progress reports Biol. Brokopondo Res. Project.*
DONSELAAR, J. VAN, 1968. Water and marsh plants in the artificial Brokopondo Lake ... during the first three years of its existence. *Acta bot neerl. 17*, p. 183–196.
DONSELAAR, J. VAN, 1969. On the distribution and ecology of Ceratopteris in Surinam. *Amer. Fern. J. 59*, p. 3–8.
EIGENMANN, C. N., 1912. The freshwater fishes of British Guiana. *Mem. Carnegie Mus. 5*, 578 pp.
ENTZ, B., 1969. Limnological conditions in Volta Lake, the greatest man-made lake of Africa. *Nature and Resources 5 (4)*, p. 9–16.
FITTKAU, E. J., 1964. Remarks on limnology of Central-Amazon rain-forest streams. *Verh. Int. Ver. Limn. 15*, p. 1092–1096.
FÖRSTER, K., 1964. Desmidiaceen aus Brasilien. II. *Hydrobiologica 23*, p. 321–505.
GESSNER, F., 1955 & 1959. *Hydrobotanik I & II.* Berlin, 517 & 701 pp.
GESSNER, F., 1964. The limnology of tropical rivers. *Verh. Int. Ver. Limn. 15*, p. 1090–1092.
GESSNER, F., 1965. Zur limnologie des unteren Orinoco. *Int. Revue ges. Hydrobiol. 50*, p. 305–333.

GEISLER, R., 1967. Zur limnochemie des Igarapé Préto. (Oberes Amazomasgebiet.) *Amazoniana I*, p. 117–123.

GEISLER, R., 1969. Untersuchungen über den Sauerstoffgehalt, den biochemischen Sauerstoffbedarf und den Sauerstoffverbrauch von Fischen in einem tropischen Schwarzwasser (Rio Negro ...). *Archiv Hydrobiol. 66*, p. 307–325.

GEIJSKES, D. C., 1942. Observations on temperature in a tropical river. *Ecology 23*, p. 106–110.

GEIJSKES, D. C., 1948. Luchtfotografie en zwampbegroeiing in Suriname. *Tijdschr. K. Ned. Aardr. Gen. 65*, p. 665–668.

GEIJSKES, D. C., 1957. The zoological exploration of Suriname. *Studies fauna Suriname I*, p. 1–12.

GEIJSKES, D. C., 1967. Natuurwetenschappelijk onderzoek van Suriname: 1945– 1965. Zoölogisch onderzoek van Suriname: 1955–1965. *Nieuwe West-Indische Gids 46*, p. 12–26 & 49–58. [*Twintig jaren Studiekring*, Publ. Found. Sci. Res. Suriname and N.A. 47.]

GEIJSKES, D. C. & PAIN, T., 1957. Suriname freshwater snails of the genus Pomacea. *Studies fauna Suriname I*, p. 41–48.

GILLARD, A., 1967. Rotifères de l'Amazonie. *Bull. Inst. R. Sci. Nat. Belg. 43 (30)*, 20 pp.

GRÖNBLAD, R., 1945. De algis brasiliensibus. *Act. Soc. Sci. Fennic. (n.s.) B 2 (6)*, 43 pp.

HAMMER, L., 1965. Photosynthese und Primärproduktion im Rio Negro. *Int. Revue ges. Hydrobiol. 50*, p. 335–339.

HARDING, D., 1961. Limnological trends in Lake Kariba. *Nature 4784*, p. 119–121.

HARROY, J. P., 1954. Hydroelectricity and protection of nature. *Procès-Verbaux et Rapports. 3rd Techn. Meeting UIPN, Caracas*, p. 267–275.

HAUER, J., 1965. Zur Rotatorienfauna des Amazonasgebietes. *Int. Revue ges. Hydrobiol. 50*, p. 341–390.

HEIDE, J. VAN DER, 1967. Hydrobiology of the artificial Brokopondo Lake, Surinam. *WOTRO Report for the year 1966*, p. 46–49.

HEIDE, J. VAN DER, 1965–1967. In: *Progress reports Biol. Brokopondo Res. Project, I*, p. 65–80; II, p. 94–124; III, p. 143–160.

HEIDE, J. VAN DER, 1966. Het stuwmeeronderzoek in Suriname. *Tijdschr. Kon. Ned. Aardr. Gen. 83*, p. 173–180.

HEIDE, J. VAN DER, 1973. Plankton development during the first years of inundation of the Van Blommestein (Brokopondo) Reservoir in Suriname, S. America. *Verh. Intern. Verein Limnol. 18* (Leningrad, Dezember 1973), p. 1784–1791.

HEIDE, J. VAN DER, Hydrobiology. In: *Brokopondo Research Report*. In press.

HOEDEMAN, J. J., 1962. Voor en tegen van het Brokopondo-stuwmeer. *Elseviers Maandblad De Kern 32*, p. 31–35.

HOLM, L. G. & WELDON, L. W. & BLACKBURN, R. D., 1969. Aquatic weeds. *Science 166*, 7.XI.1969, p. 699–709.

HOLTHUIS, L. B., 1959. The Crustacea Decapoda of Suriname (Dutch Guiana). *Zool. Verhand. Leiden 44*, 296 pp.

KLINGE, H. & OHLE, W., 1964. Chemical properties of rivers in the Amazonian area in relation to soil conditions. *Verh. Int. Ver. Limn. 15*, p. 1067–1076.

KRIEGER, W., 1932. Die Desmidiaceen der Deutschen Limnologischen Sunda-Expedition. *Archiv Hydrobiol. Suppl. 11*, 129 pp.

LAND, J. VAN DER, 1970. Kleine dieren uit het zoete water van Suriname. *Zool. Bijdr. Leiden 12*, 46 pp.

126

LAWSON, G. W. & BISWAS, S., e.a. 1968. A review of hydrobiological work by the Volta Basin Research Project, 1963–1968. *Univ. of Ghana Techn. Report X 25*, 5 pp.

LEENTVAAR, P., 1958. Hydrobiologie van de Bergvennen. *Twente-nummer Natuurbeschermingswerkgroep N.J.N., RIVON-Meded. 40*, p. 21–24.

LEENTVAAR, P., 1963. Resultaten van het hydrobiologisch onderzoek van oppervlaktewater in 1960. *Water 47*, p. 203–207.

LEENTVAAR, P., 1964, 1969. In: *Progress reports Biol. Brokopondo Res. Project*. I, p. 5–10; 45–50; IV, p. 249–256.

LEENTVAAR, P., 1966. The Brokopondo Lake in Surinam. *Intern. Ass. Theor. and Appl. Limn. 16*, p. 680–685.

LEENTVAAR, P., 1966. The Brokopondo Research Project, Surinam. In: *Man-made lakes*, London, p. 33–42.

LEENTVAAR, P., 1967. The artificial Brokopondo Lake in the Suriname river. Its biological implications. *Atas Simpósio Biota Amazônica 3 (Limn.)*, Brasil, p. 127–140.

LEENTVAAR, P., 1973. Further developments in Lake Brokopondo, Surinam. *Amazoniana 4*, 1, p. 1–8.

LEENTVAAR, P., 1973. Lake Brokopondo. *Geophys. Monogr. Series 17*, Man-made lakes, p. 186–196, 5 figs.

LEENTVAAR, P., 1974. Inundation of a tropical forest in Surinam (Dutch Guiana), South America. *Proc. 1st Intern. Congress Ecology 1974*, The Hague; PUDOC, Wageningen, p. 348–354.

LINDROTH, A., 1957. Abiogenic gas supersaturation of river water. *Archiv Hydrobiol. 53*, p. 539–597.

LOWE-McCONNELL, R. H. (ed.), 1966. *Man-made lakes, London*. London-New York, 218 p.

MARGALEF, R., 1961. La vida en los charcos de agua dulce en Nueva Esparta (Venezuela). *Memoria Soc. Cienc. Nat. La Salle 21*, 59, p. 75–110.

MARLIER, G., 1955. La biologie des lacs tropicaux. *Folia scient. Africae Centr. 1 (2)*, 5 pp.

MARLIER, G., 1964. Trichoptères de l'Amazonie. *K. Belg. Inst. Nat. Wet. Verh. (2) 76*, 167 pp.

MARLIER, G., 1964. Sur trois trichoptères nouveaux ... en Amérique du Sud. *Bull. Inst. R. Sci. Nat. Belg. 40 (6)*, 15 pp.

MARLIER, G., 1965. *Etude sur les lacs de l'Amazonie centrale*. Inst. Nac. Pesquisas da Amazonia, Manaus, 51 pp.

MARLIER, G., 1967. Hydrobiology in the Amazon region. *Atas Simpósio Biota Amazônica 3* (Limn.), Brasil, p. 1–7.

MARLIER, G., 1967. Ecological studies in some lakes of the Amazon Valley. *Amazoniana 1 (2)*, p. 91–115.

OBENG, L. E., 1969. *Man-made lakes. Accra*. London, 398 pp.

Man-made lakes: A selected guide to the literature. Washington, D.C., 1965, 98 pp.

NIJSSEN, H., 1970. Revision of the suriname catfishes of the genus Corydoras ... *Beaufortia 18 (230)*, 75 + 2 pp.; thesis Amsterdam.

PANERO, R. B., 1969. A dam across the Amazon. *Science J.*, Sep. 1969, p. 56–60.

PONS, L. J., 1966. Geogenese en pedogenese in de jong-holocene kustvlakte van de drie Guyanas. *Tijdschr. K. Ned. Aardr. Gen. 83*, p. 153–172.

PARAENSE, W. L. & DESLANDES, W., 1960. Drepanotrema surinamense with an addendum on D. petricolla (Planorbidae). *Rev. Brasil. Biol. 20*, 3.

Progress reports of the Biological Brokopondo Research Project, Surinam 1964–1969. I, 1963–1964, p. 1–89; II, 1965, p. 90–141; III, 1965–1966, p. 142–207; IV, 1967–1969, p. 208–265. Found. Scient. Res. Surinam and Neth. Ant., Utrecht.

ROBERTS, W. J., 1971. Man-made lakes, their problems and environmental effects. *Nature and resources 7 (4)*, p. 14–18.

ROYEN, P. VAN, 1951. *The Podostemaceae of the New World, I.* Thesis, Utrecht, 151 pp.

RUTTNER, F., 1932. Merenonderzoek in Nederlands-Indië. *Tropische Natuur 21*, p. 178–184.

RUTTNER, F., 1937. Stabilität und Umschichtung in tropischen und temperierten Seen. *Archiv Hydrobiol. Suppl. 15*, p. 178–186.

SCHMASSMANN, H., 1951. Untersuchungen über den Stoffhaushalt fliessender Gewässer. *Schweiz. Ztschr. Hydrol. 13*, p. 300–335.

SCHMASSMANN, H., 1955. Die Stoffhaushalts-Typen der Fliessgewässer. *Archiv Hydrobiol. Suppl. 22*, p. 504–509.

SCHMIDT, G. W., 1970. Numbers of bacteria and algae and their interrelations in some Amazonian waters. *Amazoniana 2*, p. 393–400.

SCHULZ, J. P., 1954. *Vergelijkend literatuuronderzoek inzake de ecologische consequenties van het "Combinatie-plan Suriname Rivier".* Found. Scient. Research Surinam and Neth. Ant., Utrecht, 124 pp.

SCOTT, A. M. & GRÖNBLAD, R., 1957. New and interesting desmids from S.E. United States. *Acta Soc. Sci. Fennica (2) N 8*, 62 pp.

SCOTT, A. M., GRÖNBLAD, R. & CROASDALE, H., 1965. Desmids from the Amazon basin, Brazil. *Acta bot. fenn. 69*, 94 pp.

SIOLI, H., 1964. General features of the limnology of Amazonia. *Verh. Int. Ver. Limn. 15*, p. 1053–1058.

SIOLI, H., 1965. Bemerkungen zur Typologie amazonischer Flüsse. *Amazoniana 1*, p. 74–83.

SIOLI, H., 1969. Oekologie im brasilianischen Amazonasgebiet. *Naturwissenschaften 56*, p. 248–255.

STERNBERG, H. O'REILLY, 1968. Man and environmental change in South America. In: *Biography and ecology of South America, I*, p. 413–445.

THOMASSON, K., 1962. Planktological notes from western North America. *Arkiv Botanik (2) 4*, p. 437–463.

TONOLLI, V., 1966. The productivity of lakes and rivers. *New Scientist 14.IV.*, p. 119–121.

WAGENAAR HUMMELINCK, P., 1956. Caribische beelden. III, Mourera fluviatilis ... *West-Indische Gids 36*, p. 125–132.

WAGENAAR HUMMELINCK, P., 1961. Het Brokopondoplan. *Vakblad voor Biologen 41*, p. 174–179.

WESTERMANN, J. H., 1956. *Een korte beschouwing over het Brokopondo-plan Suriname* ... Found. Scient. Research Surinam and Neth. Ant., Utrecht, 16 pp.

WESTERMANN, J. H., 1971. Historisch overzicht van de wording en het onderzoek van het Brokopondo-stuwmeer. *Nieuwe West-Indische Gids 48*, p. 1–55.

TABLES 1–26

T A B L E   1

PHYSICO-CHEMICAL DATA OF THE UPPER SURINAME RIVER

Electric conductivity in micro Siemens; temperature in centigrades

L o c a l i t y	date	μ S	pH	$^{\circ}$C
Pikien Rio	13-17.VIII.64	16	5.3	28.6
same locality	13-18. IX. 64	18	5.4	30.0
Gran Rio	13-17.VIII.64	16	5.3	27.5
same locality	13-18. IX. 64	19	5.5	29.0
Suriname River, Semoisie	13-17.VIII.64	17	5.3	28.5
same locality	13-18. IX. 64	18	5.4	30.5
Suriname River, Pokigron	13-17.VIII.64	19	5.2	28.5
same locality	13-18. IX. 64	19	5.5	31.0
Suriname River, Mamadam	18. I. 64	20	5.8	-
same locality	18. IX. 64	19	5.4	-
Suriname River, Grankreek	25.XII. 63	22	6.4	-
Suriname River, Kabel	18. I. 64	20	6.4	-
Suriname River, Afobaka	XI.63 - II.64	22-25	6.0-6.8	-

T A B L E   2

PHYSICO-CHEMICAL DATA OF THE GRANKREEK

Electric conductivity in micro Siemens; temperatures in centigrades

L o c a l i t y	date	μ S	pH	$^{\circ}$C
100 m upstream (depth 1.5 m; transp. 1.2 m; alkal. 0.40; 6.0 mg O_2/l)	25.XII.63	47	6.7	27.7
1-2 km upstream	5.III.64	45	6.3	-
6 km upstream	5.III.64	45	6.3	-
in a small stream	5.III.64	70	6.1	-
500 m upstream (almost stagnant; transp. 0.7 m)	15.VII.64	31	5.5	27.1
1 km upstream (stagnant)	30.VII.64	39	5.8	-
at first great rapid	30.VII.64	38	5.8	-
in a small stream	30.VII.64	36	5.8	-
at second great rapid	30.VII.64	36	5.8	-

130

T A B L E 3

PHYSICO-CHEMICAL DATA OF THE SARAMACCA RIVER

L o c a l i t y	date, 1964	μ S	pH
1 = Saramacca River, Boslanti	7.IV.	17	5.1
2 = forest streamlet at Boslanti	7.IV.	46	5.0
3 = Saramacca River, Grandam	8.IV.	17	5.1
4 = Saramacca River, Pakka Pakka	8.IV.	18	5.1
5 = Saramacca River, Paudam	9.IV.	17	5.1
6 = Saramacca River, Loekoesoekondre	9.IV.	18	5.1
7 = Saramacca River, Awarradam	4.IV.	19	5.0
8 = Saramacca River, Dramhosso	9.IV.	18	5.1
9 = Saramacca River, Apaugo	10.IV.	22	5.2
10 = Pikien Saramacca	10.IV.	36	5.3
11 = Saramacca River, Kwakoegron	10.IV.	20	5.2

T A B L E 4

PHYSICO-CHEMICAL DATA OF THE TIBITI, COPPENAME AND TAPANAHONY RIVERS

L o c a l i t y	date, 1964	μ S	pH	mg Cl/l
swamp near Tibiti	25.I.	34	4.9	9
swamp near Tibiti	26.I.	34	4.8	8
1 hour upstream from Tibiti Sabana	26.I	28	4.8	5
Tibiti near Sabana, LW, 12.00 hr	26.I.	34	5.1	9
Tibiti near Sabana, HW, 17.00 hr	26.I.	41	5.3	11
Coppename near Tibiti	26.I.	75	6.6	18
Upper Tapanahony near Paloemeu	8.III.	18	5.8	-
Tapanahony, Maboegoe falls	8.III.	17	5.8	-
Tapanahony, downstream Maboegoe falls	8.III.	16	6.1	-

T A B L E 5

PHYSICO-CHEMICAL DATA OF MAROWIJNE, COTTICA, COMMEWIJNE AND SURINAME RIVERS

L o c a l i t y	date, 1964	μ S	pH	mg Cl/l
Marowijne near Bigiston, HW	5.IX.	27	5.7	1
Marowijne near Albina, LW	6.IX.	30	5.7	1
Cottica near Moengo	6.IX.	300	3.6	35
Commewijne	6.IX.	260	3.8	60
Suriname River near Paramaribo	6.IX.	12800	-	4600

T A B L E 6

PHYSICO-CHEMICAL DATA OF RIVULETS, SMALL STREAMS AND RIVERS

L o c a l i t y	date	μ S	pH	mg O_2/l	°C	colour
r i v u l e t s						
near road Afobaka-Brownsweg, A	13. XI.63	45	5.0	6.6	27	colourless
(alkal. 0.25)						
same locality	26. XI.63	35	5.4	2.6	24	colourless
near road Afobaka-Brownsweg, B	26. XI.63	45	5.9	4.9	24	colourless
near road Afobaka-Brownsweg, C	26. XI.63	33	5.7	3.3	24	turbid
near road Afobaka-Brownsweg, D	26. XI.63	82	6.8	2.6	24	colourless
well in quarry at Afobaka	9. XII.63	175	7.9	-	-	iron-brown
near Assidonhoppo	13.VIII.64	24	5.3	-	25	colourless
near Ligolio, 1	15.VIII.64	18	5.2	-	24.2	colourless
near Ligolio, 2	15.VIII.64	18	5.2	-	24.5	colourless
s m a l l s t r e a m s						
Carolina kreek near Zanderij	16. XI.63	35	4.0	2.6	25	dark brown
same locality in savanna	7. XII.63	39	3.6	-	-	dark brown
same locality	5. VII.64	70	3.3	-	-	dark brown
Coropina kreek near Republiek	28. VI.64	62	3.3	-	-	dark brown
same locality	5. VII.64	49	3.4	-	-	dark brown
Compagnie kreek near Afoboka road	20. XI.63	32	6.1	3.8	25	colourless
Wedang kreek	20. XI.63	33	5.8	-	-	colourless
Witte kreek S of Kabel	18. II.64	110	7.2	-	24.2	colourless
2nd creek S of Kabel	18. II.64	38	6.1	-	24.5	colourless
Kassie kreek near Pokigron	20. III.64	23	6.4	-	-	colourless
Anjanwoyo kreek near Botopassie	16.VIII.64	25	5.2	-	26.2	turbid, brown
same locality	16. IX.64	27	5.4	-	30	turbid, brown
Sipari kreek near Aurora	17.VIII.64	19	5.3	-	26	turbid, brown
same locality	17. IX.64	19	5.4	-	27	turbid, brown
Macambi kreek near Brownsweg	26. VII.64	90	6.2	-	-	colourless
r i v e r s						
Suriname River: Table 1, Fig. 10						
Sara kreek: Fig. 10						
Locus kreek	25. XII.63	70	-	-	25	-
Grankreek (alkalinity 0.40)	25. XII.63	47	6.7	6.0	27.7	yellow brown
Tapanahony, Paloemeu	8. III.64	18	5.8	-	-	colourless
Tapanahony, Paloemeu, Maboega falls	8. III.64	17	5.8	-	-	colourless
Tibiti near Sabana, HW	25. I.64	41	5.3	-	-	dark brown
same locality, LW	25. I.64	34	5.1	-	-	dark brown
Tibiti 1 h. upstream Sabana	25. I.64	28	4.8	-	-	brownish
Coppename, near Tibiti	25. I.64	75	6.6	-	-	yellow green
Marowijne near Albina, LW	6. IX.64	30	5.7	-	-	very turbid
(1 mg Cl'/l)						
Marowijne near Bigiston, HW						
(1 mg Cl'/l)	5. IX.64	27	5.7	-	-	turbid,
						colourless
Cottica near Moengo (35 mg Cl'/l)	5. IX.64	300	3.6	-	-	opalescent
Commewijne near E-W road	5. IX.64	260	3.8	-	-	light brown
(60 mg Cl'/l)						
Saramacca River: Table 3						

TABLE 7

PHYSICO-CHEMICAL DATA OF FOREST SWAMPS, GRASS SWAMPS AND POOLS

Locality	date	μ S	pH	mg O_2/l	^{o}C	colour
forest swamp at Afobaka	19. XI.63	52	6.3	-	-	yellow
same locality	22. XI.63	51	6.2	-	30.5	yellow
grass swamp at Afobaka	22. XI.63	60	6.6	-	35.0	green
same locality	17.XII.63	57	6.2	-	-	green
same locality	9. I.64	68	6.4	3.4	-	green
swamp with tree stumps	19. XI.63	70	6.6	-	-	green
same locality at Afobaka	22. XI.63	85	6.0	-	33.5	yellow
swamp with tree stumps and *Typha*, Afobaka road	20. XI.63	35	5.8	6.0	28.5	yellow
pool near dam at Afobaka, no vegetation	19. XI.63	120	7.2	-	-	clear
pool near dam at Afobaka, with vegetation	19. XI.63	79	7.3	-	-	yellow
pool near 4th dam near Afobaka, no vegetation	19. XI.63	210	7.1	7.7	30.0	clear
same locality, effluent (alkalinity 1.0)	25. XI.63	225	7.2	-	30.0	clear
pool near 4th dam, with dead trees (alk. 1.0)	19. XI.63	100	6.4	1.9	30.0	dark yellow
same locality	25. XI.63	102	6.2	2.0	30.0	dark yellow
pool with dead trees on road to 4th dam, A	25. XI.63	60	6.0	3.5	27.5	yellow
same locality, B	17.XII.63	54	6.5	-	-	yellow
pool with dead trees on road to 4th dam, B	25. XI.63	80	6.4	3.2	29.5	grey
same locality	9. XI.63	65	6.4	5.6	-	grey
pool with dead trees on road to 4th dam, C	25. XI.63	62	6.6	-	-	grey
rainwater at Afobaka	14. XI.63	7	4.1	-	-	colourless
the same	9.XII.63	7	4.5	-	-	colourless
oxydation pond at Afobaka	9. I.64	180	6.6	0	-	dark green
grass swamp at Coropina, near Republiek	28. VI.64	28	5.7	-	-	-
old branch of Surinam River, near Republiek	30. I.64	20	5.1	-	-	-

T A B L E 8

ELECTRIC CONDUCTIVITY IN BROKOPONDO LAKE

at different depths near Afobaka, Sta.1, 1964 (in μ S)

Depth	20.VII.	21.VII.	24.VII.
0 m	28	26	27
1.5 m	28	27	27
2.5 m	35	34	35
3.5 m	30	28	28
4.5 m	31	29	29
5.5 m	31	29	29

T A B L E 9

ELECTRIC CONDUCTIVITY IN BROKOPONDO LAKE

at different localities, 1964 (in μ S)

L o c a l i t y	16.VI.	17.VI.	19.VI.	21.VI.	24.VI.	2.VII.	7.VII.
Koffiekamp, left bank	25	24	24	24	24	24	24
Koffiekamp, midst	26	24	25	26	26	24	26
Koffiekamp, right bank	27	26	26	25	25	26	25
Afobaka, in the middle	26	24	25	25	25	25	24
Afobaka, in the forest	27	25	26	25	26	25	24
Saddle dam, East	26	25	26	25	25	25	24
bottle with filamentous algae	72	150	225	250	275	300	290

134

T A B L E 10

OBSERVATIONS IN A LENGTH SECTION OF THE LAKE, 29/30.IV.1964

Locality	Beerdotti	Geltoesi	Gansee	Lombe	Saidagoe	Kabel-S	Kabel	Kadjoe	Aloesoebanja -S	Aloesoebanja -M	Koffiekamp	Afobaka
colour	brown	brown	brown	brown	green	green	green	flocky brown	flocky brown	flocky brown	flocky brown	flocky brown
°C	30.3	31.2	31.0	32.0	32.0	30.3	29.8	-	-	30.4	-	31.2
mg O_2/l	7.1	7.3	6.8	6.9	6.9	4.5	2.6	-	-	2.5	-	4.6
detritus	much	few	few	scarce	scarce	iron bact.	-	iron bact.	iron bact.	iron bact.	flocky matter	flocky matter

T A B L E 11

OXYGEN CONTENT AT AFOBAKA, 17.IV.1964

open water and amongst trees

left bank, about 30 m in the forest	1.8 mg/l
left bank, near border of former river bed	2.9 mg/l
in the middle of former river	3.2 mg/l
right bank, near border of former river bed	3.4 mg/l
right bank, about 50 m in the forest	4.0 mg/l

T A B L E 12

OXYGEN CONTENT IN FORMER SARA KREEK, 17.IV.1964

open water near former Suriname River	3.7 mg/l
in the forest, open space	3.8 mg/l
500 m further, open water	4.1 mg/l
500 m further, open water	2.8 mg/l
500 m further, open water	2.0 mg/l
500 m further, open water	2.3 mg/l
1000 m further, open water	2.5 mg/l
1500 m further, open water	2.7 mg/l
2500 m further, open water	1.7 mg/l
same station, right bank, in forest, open, much filamentous algae	2.2 mg/l
same station, left bank, in forest, shady, few filamentous algae	1.5 mg/l
1000 m further, station Sara, open water	1.8 mg/l

TABLE 13

DECREASE OF OXYGEN IN BOD-BOTTLES AT AFOBAKA SAMPLED 26.XI.1963

Locality	26.XI.	29.XI. dark	29.XI. light
Suriname River	7.0	4.3	5.8
forest swamp	4.8	2.0	3.6
open swamp	8.0	3.8	5.3
oxydation pond	9.1	0.0	2.8

TABLE 14

DECREASE OF OXYGEN IN BOD-BOTTLES SAMPLED 9.I.1964

(O_2 in mg/1; bottles kept in the dark)

Date	Suriname River	Sara Kreek	forest swamp	open swamp
9.I.	6.4	6.0	5.6	3.4
10.I.	6.6	5.2	5.0	2.4
11.I.	6.4	5.0	4.8	2.8
12.I.	6.3	5.0	4.1	2.6
14.I.	6.1	5.3	0.4	0.9
16.I.	4.9	5.4	1.8	0.3

TABLE 15

DECREASE OF OXYGEN IN BOD-BOTTLES AT AFOBAKA SAMPLED 15.I.1964

(O_2 in mg/1; bottles kept in the dark)

Depth	Suriname River, 15.I.	after 3 days
0 m	7.1	7.3
1.5 m	7.4	6.2
2.5 m	7.3	6.7
4.0 m	7.7	6.4
5.5 m	7.7	5.5

TABLE 16

DECREASE OF OXYGEN IN BOD-BOTTLES AT AFOBAKA IN THE TRANSITION PERIOD, 1964

(O_2 in mg/l; bottles kept in the dark)

Depth	25.III.	28.III.	1.IV.	4.IV.	8.IV.	11.IV.
0 m	1.9	1.0	2.9	2.1	4.2	2.8
1.5 m	1.9	0.9	2.7	2.0	2.2	1.0
2.5 m	1.2	0.6	2.1	1.6	0.4	0.0
4.5 m	0.0	0.0	0.0	0.0	0.0	0.0

TABLE 17

DECREASE OF OXYGEN IN BOD-BOTTLES AT AFOBAKA AFTER RESTABILIZATION, 1964

(O_2 in mg/l)

Depth	20.V.	23.V. dark	23.V. light	24.VIII.		27.VIII. dark
0 m	3.9	0.0	0.0	6.6		2.5
1 m	3.4	0.0	0.0	5.8		1.3
1.5 m	2.8	0.0	0.0	-		-
2 m	-	-	-	2.9		0.0
2.5 m	0.0	-	-	-		-
3 m	-	-	-	0.0		-

Depth	24.V.		26.V.	27.V.	28.V.	29.V.	30.V.
0 m	7.3	dark	4.0	2.7	1.9	0.1	0.1
		light	5.3	4.6	5.0	7.6	6.6

T A B L E 18

DECREASE OF OXYGEN IN BOD-BOTTLES BETWEEN AFOBAKA AND KABEL, 1964

(O_2 in mg/l; bottles kept in the dark)

Date	Afobaka	Koffiekamp	Aloesobanja N.	Aloesobanja S.	Kadjoe	Kabel
12.III.	1.5	2.3	4.8	7.7	8.5	9.2
after 3 days	1.5	1.4	4.7	4.8	5.3	9.1
18.III.	1.0	1.1	3.7	5.6	6.5	7.4
after 3 days	-	0.6	3.3	4.7	6.1	7.0
25.III.	1.9	4.0	6.5	7.7	8.6	7.5
after 3 days	1.0	4.6	5.3	6.6	7.2	6.6
1.IV.	2.9	3.0	3.7	4.7	5.4	6.5
after 3 days	2.1	2.4	3.3	3.9	4.2	5.4

T A B L E 19

DECREASE OF OXYGEN IN BOD-BOTTLES AT KABEL, 1964

(O_2 in mg/l; bottles kept in the dark)

Depth	9.II.	26.II.	4.III.	11.III.	18.III.	25.III.	1.IV.	8.IV.
0 m	9.0	8.4	8.2	10.0	7.4	7.5	6.5	6.1
after 3 days	6.5	6.5	6.5	6.0	7.0	6.6	5.4	4.0
1.5 m	10.2	9.2	7.4	9.7	7.5	7.5	5.4	6.1
after 3 days	6.5	6.7	6.5	6.0	6.1	6.9	4.7	3.5
2.5 m	10.8	7.9	8.1	8.2	7.0	8.0	4.4	2.4
after 3 days	6.5	6.8	6.4	6.0	6.2	6.9	4.0	1.4
4.5 m	8.8	7.7	7.0	6.9	7.2	6.6	3.7	2.2
after 3 days	7.0	6.8	6.5	5.6	6.3	6.9	3.2	1.4
6.5 m	7.5	7.6	8.3	7.3	7.2	6.7	2.9	0.0
after 3 days	5.8	6.9	6.2	6.0	6.4	7.3	2.5	0.0

TABLE 20

DECREASE OF OXYGEN IN BOD-BOTTLES AT KABEL AFTER RESTABILIZATION, 1964

(O_2 in mg/l)

Depth	20.V.	after 3 days dark	light	24.VIII.	after 3 days dark
0 m	1.7	0.0	0.0	3.9	1.0
1 m	0.1	0.0	0.0	2.3	1.0
2 m	0.0	-	-	1.4	0.0
3 m	-	-	-	0.0	-

TABLE 21

CHEMICAL DATA

(Analyzed by the Centraal Laboratorium, Paramaribo)

1963	Sara 4.XII. 0 m	Afobaka 7.XII. 0 m
Cl mg/l	5	3
Ca mg/l	4.5	2.6
$KMnO_4$ mg/l	15.0	13.5
HCO_3 hardness	1.5	0.7
total hardness °D	1.3	0.4
pH	6.9	7.1

NO_3, NO_2, NH_4, PO_4 and SO_4 not detectable.

1.IV.1964	Afobaka 0 m	Afobaka[+] 6.5 m	Kadjoe[+] 0 m	Kabel 0 m	Locus 0 m
Fe mg/l	0.2	0.4	1.0	1.0	1.0
$KMnO_4$ mg/l	1.6	2.1	2.3	2.0	2.1
total hardness °D	0.45	0.60	0.45	0.5	0.9
pH	5.6	5.6	5.6	5.6	5.6

PO_4 all strongly positive; Cl, NO_2, NH_4, SO_4 and H_2S not detectable; samples colourless and scentless.

[+]Analyses from 'Sunevo', Paramaribo:
Afobaka, 6.5 m: PO_4 2.4 mg/l, SiO_2 11.0 mg/l, NO_3 0.5 mg/l.
Kadjoe, 0 m: NO_3 1.0 mg/l.

T A B L E 22

CHEMICAL DATA

(Analized by the Centraal laboratorium, Paramaribo)

30.IV.1964	Afobaka 0 m	Afobaka 17.5 m	Beerdotti 0 m	Saidagoe 0 m	Saidagoe 5.6 m
PO_4 mg/l	-	-	1.0	1.0	1.0
NO_3 mg/l	0.03	0.06	0.05	0.13	0.10
Fe mg/l	0.3	0.8	0.2	0.7	0.9
$KMnO_4$ mg/l	-	-	13.7	-	-

AFOBAKA 20.V.1964

Depth in m	0	1.5	2.5	4.5	6.5	8.5	10.5	12.5	14.5	16.5
Fe mg/l	0.69	0.64	0.66	1.23	0.66	1.00	1.02	1.00	1.45	1.82
NO_3 mg/l	0.5	1.5	2.0	0.5	0.5	1.0	1.5	1.0	1.0	1.2
PO_4 mg/l	0.48	0.44	0.40	0.60	0.48	0.82	0.60	0.66	1.00	1.08
SO_4 mg/l	0.0	-	-	-	-	-	-	-	-	0.0

AFOBAKA 3.VIII.1964

Depth in m	0	1.5	2.5	4.5	6.5	8.5	10.5	12.5	14.5	16.5	18.5	20.5	22.5	24.5
NO_3 mg/l	2.0	1.5	1.0	1.5	2.0	0.0	1.5	0.5	1.3	1.5	1.3	0.5	0.5	1.3
PO_4 mg/l	0.7	0.8	0.8	0.8	1.0	0.8	0.8	0.8	0.8	1.2	0.8	16.8	1.5	1.0
SO_4 mg/l	0.0	0.0	0.0	0.0	0.0	0.0	0.0	0.0	0.0	0.0	0.0	5.0	0.0	1.0
$KMnO_4$ mg/l	27.3	28.9	-	30.7	27.3	25.3	28.8	30.0	30.7	28.1	31.0	29.1	27.9	27.9

T A B L E 23

PHYSICO-CHEMICAL DATA OF BROKOPONDO LAKE AND SURINAME RIVER, April 1968

t   = temperature in ^{o}C

Cl = chlorid content in mg/l

K   = electric conductivity in μ S (micro Siemens)

O_2 = oxygen content in mg/l

D   = oxygen in BOD-bottles kept in the dark; D_1, D_2, D_3 = after 1, 2, and 3 days.

L   = oxygen in BOD-bottles kept in light; L_1, L_2, L_3 = after 1, 2, and 3 days.

---

9 April   AFOBAKA, 10 h.   13.45 h.

0 m	O_2= 3.8	20 m	O_2= 7.8	t = 29.0	
3 m	4.2	22 m	5.8	29.2	
5 m	3.7	24 m	7.0	29.0	
7 m	4.2				
9 m	4.6				
11 m	5.2				
13 m	5.2				
15 m	6.4				
17 m	5.1				

11 April   AFOBAKA, 10.30 h. (cloudy)

0 m	t = 29.1	O_2 = 4.7	D_1 = 3.6	D_3 = 1.2
5 m	28.6	4.8	3.6	2.5
10 m	28.2	4.8	2.2	0.1
15 m	27.8	7.0	5.8	0.3
20 m	27.6	7.0	5.4	0
25 m	27.2	5.0	3.2	0
30 m	27.1	2.6	-	-
35 m	27.0	3.2	-	-
36 m	26.9	3.4	1.2	-

KABEL, 13.15 h. (sun)

0 m	t = 30.2	O_2 = 6.8	D_1 = 5.6	D_3 = 4.0
5 m	29.6	7.1	5.7	4.7
10 m	28.8	5.2	1.8	0.6
15 m	28.0	6.1	4.6	2.2
20 m	28.2	7.4	6.0	3.2
25 m	27.8	7.3	4.6	2.8

SARA, 12 h. (cloudy)

0 m	t = 29.7	O_2 = 7.5	D_1 = 5.2	D_3 = 5.6
5 m	29.0	6.6	5.8	4.9
10 m	28.5	5.2	0.8	0.3
15 m	27.9	6.5	5.0	0.4
20 m	27.3	7.8	6.5	0.3
25 m	27.0	8.8	6.4	0.2
30 m	27.1	5.2	2.0	0

RAILROAD POINT 3, 15 h. (cloudy)

0 m	t = 31.0	O_2 = 4.4	D_1 = 2.9	D_3 = 0.9
2 m	29.4	3.5	2.6	2.7

T A B L E 23 (continued)

12 April   LOCUS, 10.30 h. (sun)

		O_2	D_2	pH	K
0 m	t = 31.8	= 6.0	= 3.4	= 6.7	= 70
5 m	29.7	6.0	3.0	6.5	70
10 m	28.9	4.6	0	6.0	77
15 m	28.0	5.2	0.1	5.9	99
20 m	27.6	3.6	0	5.9	94
24 m	27.2	7.5	0	5.9	115

SIKAKOEMPOE, 13 h. (many Eichhornia, algal mats, no wind)

		O_2	D_2	pH	K
0 m	t = 33.8	= 3.3	= 1.8	= 6.2	= 54
5 m	29.5	2.1	0	6.0	62
8 m	29.2	2.8	0	5.8	74

14 April   PHEDRA, 15 h.

$$O_2 = 0.6$$

BERG EN DAL, 16 h.

$$O_2 = 0.8$$

15 April   GRANKREEK, 11.30 h. (sun)

		O_2	L_2	D_2	pH	K	Cl
0 m	t = 32.0	= 5.7	= 5.4	= 4.2	= 6.6	= 26	= 5
5 m	29.3	6.1	6.6	5.3	6.2	27	6
10 m	29.0	4.8	4.0	4.2	6.0	30	6
15 m	29.0	6.5	3.9	3.4	5.9	31	6
17 m	28.2	6.2	-	-	5.9	31	6

BEDOTTI, 14 h. (sun)

		O_2	L_2	D_2	pH	K	Cl
0 m	t = 33.1	= 5.8	= 3.4	= 2.8	= 6.4	= 40	= 8
5 m	29.5	4.8	3.9	4.0	6.2	32	6
10 m	28.7	4.8	-	1.0	6.0	39	8
15 m	28.1	5.6	1.4	0	5.8	43	9
18 m	28.5	2.0	1.0	0	5.9	36	7

16 April   AFOBAKA, 15.30 h. (wind, clouds)

		O_2	L_3	D_3	pH	K	Cl
0 m	t = 30.8	= 7.0	= 5.6	= 5.0	= 6.9	= 39	= 8
5 m	28.8	5.9	5.3	4.8	6.5	38	8
10 m	28.2	5.0	0	0	5.9	42	8
15 m	28.0	5.3	0	0	5.9	39	8
20 m	27.6	6.4	0	0	5.9	45	9
25 m	27.3	6.7	0	0	6.0	65	16
30 m	27.2	1.2	0	0	6.0	120	29
35 m	27.9	1.6	0	0	6.0	140	31

17 April   SARA, 14 h. (sun)

		O_2	L_3	D_3	pH	K	Cl
0 m	t = 31.2	= 6.9	= 6.4	= 3.1	= 7.0	= 38	= 8
5 m	29.3	8.2	7.3	-	6.8	38	8
10 m	28.7	6.9	0	0	6.3	49	10
15 m	28.1	6.5	0	0	6.0	60	15
20 m	27.3	6.6	0	0	5.9	80	20
25 m	27.7	6.1	0	0	5.8	80	20
30 m	27.5	2.7	0	0	5.9	90	20

T A B L E 23 (continued)

18 April   BAY AUXILIAIR DAM 4 (sun)

0 m	t = 31.5	O_2 = 7.6	
2.5 m	28.8	5.2	
5 m	28.1	0	
10 m	27.9	0	
15 m	26.9	0	

19 April   KABEL, 15.45 h. (cloudy)

0 m	O_2 = 6.9	L_2 = 6.7	D_2 = 6.4
2.5 m	5.8	6.0	6.6
5 m	5.4	6.0	5.3
10 m	6.6	0	0
15 m	7.9	5.5	4.3
20 m	0	0	0
23 m	0	0	0

23 April POKIGRON, 4.40 h. (cloudy, rain storms 16 - 17 h., gauge 4.40 m, high water)

	8 h.	10 h.	12 h.	14 h.	16 h.	18 h.	20 h.	
0 m	t = 25.9	26.0	26.2	27.0	26.9	26.4	26.4	pH = 5.6
2.5 m	25.8	26.5	26.3	27.1	27.0	26.3	26.2	
0 m	O_2= 6.4 .	6.3	6.6	6.2	6.8	6.0	5.9	K = 21
2.5 m	7.1	6.7	6.5	6.8	6.4	6.6	6.1	

24 April MAMMADAM, 9.45 h.

0 m	t = 26.4	O_2 = 7.1	D_4 = 5.8	pH = 5.5	K = 22

25 April MAMMADAM, 9 h.

0 m	t = 26.5	O_2 = 7.3	D_3 = 6.1
9.5 m	26.5	7.5	

ADEWAI, 9.30 h. (plankton brown)

0 m	t = 27.3	O_2 = 6.3	D_3 = 5.8	pH = 5.6	K = 21
2.5 m	27.0	-			
5 m	27.0	7.7			
10 m	26.9	7.4			

PANPANTIRI, 10.15 h. (plankton green)

0 m	t = 29.4	O_2 = 11.1	D_3 = 5.3	pH = 5.6	K = 22
5 m	27.1	6.8	5.8		
15 m	26.8	10.2	7.2		

27 April REDIDOTTI, 11 h.

0 m	t = 30.8	O_2 = 3.6	D_2 = 1.6	pH = 5.7	K = 36
2.5 m	29.2	3.3	2.3		
5 m	29.1	3.5	2.4		
10 m	28.0	5.1	1.6		

STATION 8, 12 h.

0 m	t = 31.1	O_2 = 6.1	D_2 = 2.7	pH = 6.2	K = 38
5 m	29.8	3.0	-		
10 m	28.7	4.9	0.9		
15 m	28.0	3.7	0		

T A B L E 24

DAILY FLUCTUATIONS OF TEMERATURE AND OXYGEN AT SARA, 29.IV.1968

+ = living crustaceans
O_2= oxygen content in mg/l (+)= no living crustaceans observed
− = no crustaceans observed

Depth in m	8.15h (sun) °C	O_2	10.30h (sun) °C	O_2	12.30h (sun) °C	O_2	14.30h (sun) °C	O_2	16.45h (sun) °C	O_2	18.45h (dark) °C	O_2
0	30.0	5.7	31.2	6.7 +	31.4	7.1 +	32.5	7.0 +	31.7	7.2 +	30.5	7.1 −
2.5	30.0	6.5	31.5	6.5 +	31.0	7.0 +	31.6	7.2 +	30.3	7.1 +	28.9	6.0 −
5	29.2	0.4	30.3	3.1 +	30.5	4.0 +	31.0	4.6 +	29.7	4.6 +	28.0	4.4 −
10	28.5	0	29.5	0 ?	29.4	3.9 +	29.9	4.8 +	28.7	5.5 ?	28.1	4.8 −
15	28.3	0.3	28.9	0 ?	28.6	4.9 +	29.0	6.8 +	28.1	7.1 ?	27.5	5.9 −
20	29.3	3.2	28.3	0 −	28.4	5.1 ?	29.0	7.0 +	27.8	8.6 ?	27.2	5.9 −
25	29.8	5.8	30.1	5.1 +	28.0	9.6 ?	28.4	7.0 ?	27.4	3.7 (+)	27.0	4.0−−
30	27.7	0					27.9	2.6 ?	27.3	1.6	26.6	1.8 −

T A B L E 25

DAILY FLUCTUATIONS OF TEMPERATURE AND OXYGEN AT AFOBAKA, 21.IV.1968

O_2= oxygen content in mg/l   + = living crustaceans
(+)= no living crustaceans observed
$\underline{f}$ = gas bubbles and H_2S − = no crustaceans observed

Depth in m	8 h (rain) °C	O_2	10 h (rain) °C	O_2	12 h (rain) °C	O_2	14 h (sun) °C	O_2	16 h (cloudy) °C	O_2	18 h (rain) °C	O_2
0	28.5	5.9 +	28.2	7.0 +	28.3	6.3 +	29.0	6.5 +	29.4	6.9 +	28.5	6.9 +
2.5	28.5	5.8 +	28.2	6.3 +	28.3	6.1 +	29.1	6.0 +	29.0	6.9 +	28.7	6.1 +
5	28.1	0.9 +	28.0	1.1 +	28.2	1.5 +	28.9	4.8 +	29.4	6.5 +	28.4	5.7 +
10	27.7	0.3 −	27.6	0.8 −	28.5	6.0 +	28.2	1.1 +	27.9	0.7 +	27.6	0 +
15	27.0	0.5 −	28.0	1.0 +	27.0	1.0 +	27.5	0.9 +	27.2	0.7 +	28.2	5.9 +
20	26.6	0 −	28.1	6.1 −	28.0	1.2 +	27.3	0.4 +	28.5	5.4 +	26.6	0 +
25	27.2	0.7 +	28.0	6.4 +	28.1	6.1 +	26.8	0 $\underline{f}$ −	28.0	1.5 +	28.1	6.1 +
30	27.5	5.8 +	27.9	6.2 +	26.0	0 $\underline{f}$(+)	26.6	0 $\underline{f}$ −	26.1	0 $\underline{f}$ −	26.0	0 $\underline{f}$ −
35	27.4	0.7 +	25.8	0 $\underline{f}$+	25.9	0 $\underline{f}$(+)	26.6	0 $\underline{f}$ −	26.1	0 $\underline{f}$ −	26.4	0 $\underline{f}$(+)

T A B L E 26

PHYSICO-CHEMICAL OBSERVATIONS AT SARA, LOCUS, BEDOTTI, KABEL AND AFOBAKA, 30.IV.1968

O_2 = oxygen content in mg/l
D_3 = oxygen in BOD-bottles kept in the dark after 3 days
K = electric conductivity in μ S (micro Siemens)
S = Secchi-disc in m

	SARA, 10.30 h (sun)					LOCUS, 11.30 h (sun)				BEDOTTI, 13.15 h (sun)				
	$^{\circ}C$	O_2	D_3	pH	K	$^{\circ}C$	O_2	pH	K	$^{\circ}C$	O_2	D_3	pH	K
0 m	30.5	7.6	2.7	6.2	35	31.2	7.0	6.3	40	32.0	7.5	2.0	5.8	27
5 m	30.3	4.4	4.0	6.0	34	30.7	6.7	· 6.3	38	30.0	3.9	3.0	5.8	26
10 m	30.0	4.9	-	5.9	44	29.3	5.9	5.9	46	29.3	5.8	2.0	5.7	26
15 m	29.3	5.4	0	5.8	60	28.5	6.5	5.7	70	28.6	5.8	3.9	5.6	25
20 m	29.0	7.7	0	5.8	65	28.8	8.0	5.7	78	28.3	7.1	5.0	5.5	23
25 m	28.5	7.8	0	5.8	80									
30 m	28.2	-	0	5.8	80									
	S = 3.25					S = 3.00				S = 2.20				

	KABEL, 15.45 h (cloudy)				AFOBAKA, 16.45 h (rain)				
	$^{\circ}C$	O_2	pH	K	$^{\circ}C$	O_2	D_3	pH	K
0 m	32.0	7.3	6.4	33	30.5	6.5	4.9	6.2	35
5 m	30.2	7.2	6.2	33	28.4	3.7	2.3	6.2	37
10 m	29.1	4.6	5.8	32	28.0	4.9	0	6.1	39
15 m	28.3	6.6	5.6	34	27.4	4.9	0	5.8	35
20 m	28.1	7.4	5.6	33	27.0	9.6	0	5.9	43
25 m					26.9	5.3	0	5.9	86
30 m					26.6	-	0	6.0	100
	S = 3.00				S = 2.00				

TABLES I–XVII

Table 1. Plankton from Pokigron, Station 9 (1963/64)

month	December		January		February				March				April					May				June				July					August				September		
day	25	31	13	28	4	11	18	25	3	11	18	25	1	8	15	22	28	6	12	20	27	2	9	16	25	1	8	18	21	30	4	12	19	26	2	9	16
C r u s t a c e a																																					
Alona	-	-	-	-	-	-	-	-	-	-	-	-	-	-	-	-	-	1	-	-	-	-	-	-	-	-	-	-	-	-	-	-	-	-	-	-	-
Bosminopsis	1	1	1	1	-	1	1	1	1	-	1	1	-	-	1	1	1	1	1	1	-	-	-	-	-	-	-	-	-	-	-	-	-	-	-	-	-
Cyclops	1	1	1	1	-	1	1	1	1	-	1	1	-	1	1	1	1	1	1	1	-	-	-	-	-	-	-	-	-	-	-	-	-	-	-	-	-
Moina	1	-	-	-	-	-	-	-	-	-	-	-	-	-	-	-	-	1	-	-	-	-	-	-	-	-	-	-	-	-	-	-	-	-	-	-	-
R o t a t o r i a																																					
Anureopsis fissa	1	-	-	-	-	-	-	-	1	-	-	-	-	-	-	-	-	-	-	-	-	-	-	-	-	-	-	-	-	-	-	-	-	-	-	-	-
Ascomorpha saltans	-	1	1	-	1	1	-	-	-	-	-	-	-	-	-	-	-	-	-	-	-	-	-	-	-	-	-	-	-	-	-	-	-	-	-	-	-
Brachionus quadridentata	-	-	-	-	-	-	-	-	-	-	-	-	-	-	-	1	1	-	-	-	-	-	-	-	-	-	-	-	-	-	-	-	-	-	-	-	-
Collotheca mutabilis	-	-	1	1	1	-	1	1	-	1	1	-	1	-	-	2	2	1	1	1	1	-	-	-	-	-	-	-	-	-	-	-	-	-	-	-	-
Conochiloides coenobasis	-	-	-	-	-	1	-	-	-	-	-	-	-	-	1	2	2	1	2	1	1	-	-	-	-	-	-	-	-	-	-	-	-	-	1	-	1
Hexarthra insulina	1	1	1	1	-	1	1	1	1	1	1	1	-	1	1	1	1	1	1	1	1	-	-	-	-	-	-	-	-	-	-	-	-	-	-	-	-
Keratella americana	-	-	1	-	1	1	-	-	1	1	-	-	-	-	-	1	-	2	1	1	-	-	-	-	-	-	-	-	-	-	-	-	-	-	-	-	-
Lecane bulla	-	-	-	-	-	-	-	-	-	-	-	-	-	-	-	-	-	-	-	-	-	-	-	1	-	-	-	-	-	-	-	-	-	-	-	-	-
Lecane indeterm.	1	1	-	1	-	1	1	-	1	1	1	-	1	-	1	1	1	1	1	1	1	-	-	-	-	-	-	-	-	-	-	-	-	-	-	-	-
Manfredium	-	-	1	-	-	-	-	-	-	-	-	-	-	-	-	-	-	-	-	-	-	-	-	-	-	-	-	-	-	-	-	-	-	-	-	-	-
Mytilina macracantha	-	-	1	-	-	-	-	-	-	-	-	-	-	-	-	-	-	-	-	-	-	-	-	-	-	-	-	-	-	-	-	-	-	-	1	1	-
Platyias quadricornis	-	-	1	-	2	1	2	1	1	1	1	2	1	2	-	2	2	2	1	1	-	-	-	-	-	-	-	-	-	-	-	-	-	-	1	1	-
Polyarthra	-	-	-	1	1	-	1	1	-	-	1	1	-	2	1	1	2	2	1	1	-	-	-	-	-	-	-	-	-	-	-	-	-	-	-	-	-
Rotifer	-	-	-	1	1	1	1	-	-	1	1	-	1	-	1	-	-	-	-	1	-	-	-	-	-	-	-	-	-	-	-	-	-	-	-	-	-
Synchaeta	-	-	-	1	1	1	1	-	1	-	1	-	1	-	1	1	1	-	1	-	-	-	-	-	-	-	-	-	-	-	-	-	-	-	-	-	-
Testudinella patina f.trilobata	-	-	-	-	-	-	-	-	-	-	-	-	1	-	2	2	-	-	-	-	-	-	-	-	-	-	-	-	-	-	-	-	-	-	-	-	-
Trichocerca	-	-	-	-	-	-	-	-	-	-	-	-	-	1	-	1	1	-	1	-	-	-	-	-	-	-	-	-	-	-	-	-	-	-	-	1	-
P r o t o z o a																																					
Arcella	1	-	1	-	-	1	-	-	1	-	1	-	1	1	1	1	1	1	1	1	1	1	-	1	-	1	1	1	-	1	1	-	1	-	1	1	1
Difflugia	-	1	1	2	1	1	2	-	1	2	-	1	-	1	1	1	1	1	3	-	1	1	1	1	1	1	1	1	-	1	-	1	-	1	2	2	2
Heleozoa	1	1	2	1	2	1	1	-	-	1	-	1	-	1	1	1	1	-	-	2	-	2	1	1	-	1	1	1	1	1	-	1	-	1	1	2	2
Lesquereusia	-	-	2	1	1	1	1	1	1	1	1	-	1	1	1	1	1	-	2	-	1	2	1	1	-	1	1	1	-	1	-	-	-	-	-	-	-
Myriophrys paradoxa	-	-	-	-	-	-	-	-	-	-	-	-	-	-	-	-	-	-	-	-	-	-	-	-	-	-	-	-	-	-	-	-	-	-	-	-	-
F l a g e l l a t a																																					
Dinobryon	1	1	1	2	2	1	2	1	1	2	1	2	1	1	1	1	1	2	1	-	1	-	1	1	1	1	1	1	1	1	1	1	-	1	1	1	2
Eudorina elegans	1	1	1	2	1	2	3	1	1	2	1	2	1	1	1	1	1	1	1	-	3	-	1	1	1	-	1	1	1	1	1	1	1	-	1	1	2
Phacus	-	-	-	-	1	-	-	-	-	-	-	-	-	-	-	-	-	-	-	-	-	-	-	-	-	-	-	-	-	-	-	-	-	-	-	-	-
Rhipidodendron huxleyi	-	-	-	-	-	1	-	-	-	1	1	-	-	-	-	-	-	-	-	-	2	2	1	1	1	-	1	1	1	1	-	-	-	-	-	-	-
Synura uvella	-	-	-	-	-	-	-	-	-	-	-	-	-	-	-	-	-	-	-	-	-	-	-	-	-	-	-	-	-	-	-	-	-	-	-	-	-
C h l o r o p h y c e a																																					
Acanthosphaera zacharasiae	-	1	-	-	1	1	1	1	2	2	1	1	1	1	1	1	1	1	3	1	1	1	1	1	1	1	1	1	1	1	-	1	1	1	1	1	1
Dictyosphaerium pulchellum	-	1	-	-	1	1	1	1	2	2	2	1	1	1	1	1	1	1	1	1	2	2	1	1	1	1	1	1	-	1	-	1	1	1	1	1	1
Franceia ?	-	-	1	1	1	1	-	1	-	-	1	1	-	1	-	1	1	1	1	1	1	-	1	1	1	-	1	1	1	1	-	1	1	1	-	1	1
Hormidium ?	1	-	1	-	-	-	-	-	-	-	-	-	-	1	1	1	1	-	-	-	2	-	-	-	-	-	-	-	-	-	-	1	1	1	-	1	1
Kirchneriella lunaris	-	1	1	1	1	1	1	1	1	1	1	1	1	1	1	1	1	1	1	1	-	-	1	1	1	1	1	-	1	1	-	1	1	1	-	1	1
Mougeotia	1	-	1	-	-	-	-	-	-	-	-	-	-	1	-	1	1	-	-	-	-	-	-	-	-	-	-	-	-	-	-	1	1	1	-	1	1
Oocystis	1	1	1	1	1	-	1	-	1	1	1	1	1	1	1	1	1	1	1	1	1	-	-	-	-	-	-	-	-	-	-	-	-	-	-	1	1
Pediastrum duplex	1	-	1	-	1	-	-	-	-	-	-	1	-	-	-	-	1	-	-	-	-	-	-	-	-	-	-	-	-	-	-	-	-	-	-	1	1
Spirogyra	1	-	1	-	-	1	-	-	1	1	-	1	-	-	-	-	-	1	-	-	1	-	-	-	-	-	-	-	-	-	-	-	-	-	-	1	-

Table I. (continued)

D e s m i d i a c e a e

Closterium	1	-		1	-	-	-	-	-	-	-	-	1	-	-	1	-	-	-	-	1	-	-	-	-		-	-	-	-	-	-	-	-	-	-	-	-	-	-	-	-	-	-	-	1	-	
Cosmarium	-	1		-	-	-	-	-	-	-	1	-	-	-	1	-	-	-	-	-	1	-	-	-			-	-	-	-	-	-	-	-	-	-	-	-	-	-	-	-	-	-				
Cosmocladium	-	-		-	-	-	-	1	1	1	1	-	-	-	-	-	-	-	-	-	1	-	-	-			-	-	-	-	-	-	-	-	-	-	-	-	-	-	-	-	-	-				
Desmidium swartzi	-	-		-	-	-	-	-	-	1	-	-	-	-	-	-	-	-	-	-	-	-	-	-			-	-	-	-	-	-	-	-	-	-	-	-	-	-	-	-	-	-				
Euastrum	-	1		1	-	-	-	-	-	-	-	-	-	-	-	-	-	-	-	-	-	-	-	-			-	-	-	-	-	-	-	-	-	-	-	-	-	-	-	-	-	-				
Gonatozygon	1	1		-	1	1	2	1	1	2	1	3	2	1	1	-	1	2	1	1	1	2	-	-			-	-	-	-	-	-	-	-	-	-	-	-	-	-	-	-	-	-				
Hyalotheca mucosa	-	-		-	-	-	-	-	-	1	1	-	-	-	-	-	-	1	-	-	-	-	-	-			-	-	-	-	-	-	-	-	-	-	-	-	-	-	-	-	-	-				
Micrasterias arcuatus	1	-		1	-	-	1	-	-	-	1	-	-	-	-	-	1	-	-	-	-	-	-	-			-	-	-	-	-	-	-	-	-	-	-	-	-	-	-	-	-	-				
Micrasterias brasiliensis	-	-		-	-	-	-	-	-	1	-	-	-	-	-	-	-	-	-	-	-	-	-	-			-	-	-	-	-	-	-	-	-	-	-	-	-	-	-	-	-	-				
Penium	1	-		-	-	-	-	-	-	-	-	-	-	-	-	-	-	-	-	-	-	-	-	-			-	-	-	-	-	-	-	-	-	-	-	-	-	-	-	-	-	-				
Pleurotaenium nodosum	-	-		-	-	-	-	-	-	-	-	-	-	-	-	1	-	-	-	-	-	-	-	-			-	-	-	-	-	-	-	-	-	-	-	-	-	-	-	-	-	-				
Pleurotaenium indeterm.	-	-		-	-	-	-	-	1	-	-	-	-	-	-	-	1	-	-	-	-	-	-	-			-	-	-	-	-	-	-	-	-	-	-	-	-	-	-	-	-	-				
Sphaerozosma granulatum	1	-		-	-	1	-	1	1	1	2	-	-	1	1	-	1	-	1	1	-	-	-	-			-	-	-	-	-	-	-	-	-	-	-	-	-	-	-	-	-	-				
Staurastrum inaequale	-	-		-	-	-	-	-	-	-	-	-	-	-	-	1	2	2	2	-	-	-	-	-			-	-	-	-	-	-	-	-	-	-	-	-	-	-	-	-	-	-				
Staurastrum mamillatum	-	-		-	-	1	-	1	1	2	-	-	-	1	1	1	-	1	-	-	-	-	-	-			-	-	-	-	-	-	-	-	-	-	-	-	-	-	-	-	-	-				
Staurastrum trifidum	-	-		-	-	-	-	-	1	-	-	-	-	1	1	1	-	1	-	-	-	-	-	-			-	-	-	-	-	-	-	-	-	-	-	-	-	-	-	-	-	-				
Staurastrum indeterm.	-	-		-	-	1	-	-	1	-	-	-	-	1	-	-	-	-	-	-	-	-	-	-			-	-	-	-	-	-	-	-	-	-	-	-	-	-	-	-	-	-				

D i a t o m e a e

Actinocyclus	-	-		1	1	-	-	-	-	-	-	-	-	-	-	-	-	-	-	-	-	-	-	-			-	-	-																			
Diatoma	1	1		-	1	-	-	1	1	1	-	1	-	-	1	-	-	-	-	-	-	1	-	1			-	-	-																			
Eunotia asterionelloides	2	3		2	3	3	2	2	2	2	1	2	1	2	2	2	4	5	3	3	1	1	-	-			1	-	-																1	1	2	
Melosira granulata	-	1		-	1	1	1	1	-	1	-	-	-	-	-	-	1	1	1	-	1	-	-	-			-	-	-																1	1	1	
pennatae	1	1		1	1	-	-	-	-	-	-	-	-	-	-	-	-	-	-	-	-	-	-	-			-	-	-																1	1	1	
Pinnularia	1	-		-	1	-	-	-	-	-	-	-	-	-	-	-	-	-	-	-	-	-	-	-			-	-	-																-	-	-	
Rhizosolenia eriensis	-	-		-	1	3	3	1	3	-	3	2	2	-	-	1	1	3	4	3	3	1	1	-			-	-	-																-	-	-	
Surirella	1	-		1	1	1	1	1	-	1	-	1	1	-	-	1	-	1	-	1	1	-	1	-		1	-	1	-														1	-	1	1	1	
Synedra	-	-		-	-	-	-	-	-	-	-	-	-	-	-	-	-	-	1	-	-	-	-	-			-	-	-																-	-	-	

C y a n o p h y c e a e

Lyngbya	-	-		-	-	-	1	-	-	1	-	-	-	-	-	-	-	-	-	-	-	-	-	-			-	-	-																			
Merismopedia convoluta	1	-		1	1	1	-	1	1	1	-	1	1	-	-	-	-	-	-	-	-	1	-	-			-	-	-																			
Oscillatoria	-	-		1	-	-	-	-	-	-	-	-	-	-	-	-	-	-	-	-	-	-	-	-			-	-	-																			

Actinospora	-	-		-	-	-	-	-	-	-	-	-	-	-	-	-	-	-	-	-	1	-	-	-			1	-	-													-	-	-	-			
detritus	-	2		3	3	3	3	2	3	-	-	3	4	-	1	3	2	2	1	1	1	3	4	1	1	1	3	2	3	3	1	1	1	1	1	1	1	1	1	1	
Ephemerida	-	1		1	1	-	-	1	1	-	1	-	-	-	-	-	1	-	-	1	-	-	-	-			-	-	-											-	-	-		
Gastrotricha	-	1		-	-	-	-	-	-	-	-	-	-	-	-	-	-	-	-	-	-	-	-	-			-	-	-											-	-	-		
Hydracarinae	1	-		-	-	-	-	-	-	-	-	-	-	-	-	-	-	-	-	-	1	-	-	-			-	-	-											-	-	-		
Nematoda	-	-		-	-	-	-	-	-	-	-	-	-	-	-	-	-	-	-	-	-	-	-	1			-	-	-											-	-	-		
Odonata eggs	-	-		-	-	-	1	1	-	-	-	-	-	1	-	-	1	-	-	1	-	-	-	-			-	-	-											1	1		
Plumatella zooecia	-	-		-	1	-	-	-	-	-	-	-	-	-	-	-	-	-	-	-	-	-	-	-			-	-	-											-	-	-		
Podostemaceae (pieces)	1	-		1	-	1	-	-	-	-	-	-	-	-	-	-	-	-	-	-	-	-	-	-			-	-	-											-	-	-		
sponges spicula	1	1		1	-	1	1	-	1	1	-	1	1	-	1	1	1	1	1	1	1	1	1	1	1	1	1	1	1	1	1	-	1	-	-	-	-	-			-	1	1	
veliger	-	-		-	-	-	-	1	-	-	-	-	-	-	-	1	-	1	-	-	1	-	-	-			-	-	-											-	-	-		

147

Table II. Plankton from the Saramacca River (1964)

I = Boslanti, 7. IV.
II = Boslanti, little forest stream, 7. IV.
III = Grandam, 8. IV.
IV = Pakka Pakka, 8. IV.
V = Paudam, 9. IV.

VI = Loekoesoekondre, 9. IV.
VII = Awarradam, 4. IV.
VIII = Dramhosso, 9. IV.
IX = Apaugo, 10. IV.
X   = Pikien Saramacca, 10. IV.
XI = Kwakoegron, 10. IV.

	I	II	III	IV	V	VI	VII	VIII	IX	X	XI
C r u s t a c e a											
Alona	1	-	1	1	1	-	-	-	-	-	-
Bosminopsis deitersi	-	-	1	1	-	1	-	-	1	1	3
Bosmina	-	-	1	-	-	-	-	-	-	-	-
Chydorus sphaericus	-	1	-	-	-	-	-	-	-	-	-
Cyclops	1	4	1	1	-	-	-	-	-	1	1
Diaphanosoma brachyurum	-	-	1	-	-	-	-	-	-	-	-
Diaptomus	-	2	-	-	-	-	-	-	-	-	-
Harpacticidae	1	-	1	-	-	-	1	-	-	-	-
Moina	-	-	1	-	-	-	-	-	-	-	-
R o t a t o r i a											
Ascomorpha saltans	-	-	-	-	-	-	-	-	-	-	2
Brachionus falcatus	-	-	-	-	-	-	-	-	-	1	-
Cathypna	1	-	-	-	1	-	-	-	-	-	-
Filinia longiseta	1	5	-	-	1	-	-	-	-	-	-
Hexarthra insulina	-	-	1	-	1	-	-	-	-	-	1
Keratella americana	-	-	-	-	-	-	1	-	-	1	-
Lecane ludwigi	1	-	-	-	-	-	-	-	-	-	-
Lecane	1	-	1	1	1	-	-	-	-	-	-
Monostyla bulla	1	-	-	-	-	-	-	-	-	-	-
Platyias patulus	-	2	-	-	-	-	-	-	-	-	-
Polyarthra	1	2	1	1	1	2	1	-	1	-	4
Sinantherina	1	-	1	-	-	-	1	-	-	-	-
Testudinella brycei	1	-	1	-	-	-	-	-	-	1	-
Testudinella patina f. lobosa	1	-	-	-	-	-	-	1	-	-	-
Trichocerca bicarinatus	-	-	1	-	-	-	-	-	-	-	-
Trichocerca indet.	-	-	1	-	-	-	-	-	-	1	1
P r o t o z o a											
Arcella	1	1	1	1	1	1	1	1	1	1	1
Centropyxis	1	-	1	1	-	-	-	-	-	-	-
Difflugia	1	-	1	1	1	1	-	-	1	-	-
Euglypha	-	-	1	-	1	-	1	-	-	-	-
Heliozoa	-	-	-	1	1	1	1	-	-	-	-
Lesquereusia	-	-	1	1	-	-	-	-	-	-	-
Opalina ?	-	3	-	-	-	-	-	-	-	-	-
F l a g e l l a t a											
Dinobryon	-	-	-	-	-	-	-	-	-	2	-
Eudorina elegans	1	-	1	2	1	2	1	1	1	1	-
Euglena acus	-	-	-	-	-	-	-	-	-	1	-
Euglena oxyuris	-	-	1	-	-	-	-	-	-	-	-
Euglena indeterm.	-	2	-	-	-	-	-	-	-	-	-
Synura uvella	1	-	1	-	-	-	-	-	-	-	-
Trachelomonas armata	1	4	1	-	-	-	-	-	-	-	-
C h l o r o p h y c e a											
Ankistrodesmus falcatus	-	-	1	-	-	-	-	-	-	-	-
Coelastrum cambricum	-	-	1	-	-	1	-	-	-	-	-
Dictyosphaerium pulchellum	-	-	1	2	2	2	2	2	2	3	1
Franceia ?	-	-	1	1	-	-	-	1	-	1	-
Kirchneriella obesa	-	-	-	-	-	1	-	-	-	1	-
Microspora ?	-	-	1	1	-	-	-	-	-	1	-
Mougeotia	1	-	1	1	-	1	-	1	2	-	-
Pediastrum duplex	-	-	-	-	-	-	-	-	-	2	-
D e s m i d i a c e a e											
Closterium kuetzingii	-	-	-	-	1	1	1	2	3	-	-
Closterium indeterm.	1	3	1	1	-	-	-	1	1	1	1
Cosmarium spec. 1	1	-	1	-	1	-	-	1	-	-	-
Cosmarium spec. 2	-	-	-	1	1	1	1	2	2	1	-
Euastrum	-	-	-	1	-	-	-	-	-	-	-
Gonatozygon mucosum	1	-	1	-	1	1	3	2	4	3	2
Micrasterias arcuatus	1	-	1	1	-	-	-	-	-	1	-
M. mahabhalaruswensis	-	-	-	-	-	-	1	1	1	1	-
Micrasterias indeterm.	1	-	-	-	-	-	-	-	-	-	-

Table II. (Continued) Plankton from the Saramacca River (1964)

	I	II	III	IV	V	VI	VII	VIII	IX	X	XI
Pleurentaenium	1	-	-	-	-	-	-	-	-	-	1
Sphaerozosma granulatum	-	-	-	-	-	-	-	-	-	1	3
Staurastrum minnesottense	1	-	1	1	1	1	-	-	-	-	2
Staurastrum indeterm.	-	-	-	1	2	2	2	2	2	2	2
Cyanophyceae											
Oscillatoria	1	-	1	-	1	1	-	1	-	-	-
Merismopedium	-	-	-	-	-	1	-	1	-	-	-
Spirulina	-	-	-	-	-	-	-	-	1	-	-
Diatomeae											
Diatoma	1	-	1	-	1	-	1	1	2	-	-
Eunotia asterionelloides	-	-	1	1	1	1	-	1	1	1	2
Melosira granulata	1	-	2	4	3	3	3	3	1	2	1
Nitzschia	2	-	2	1	1	1	1	1	-	-	-
Pinnularia	1	-	-	1	1	-	-	-	-	1	-
Rhizosolenia eriensis	-	-	1	-	1	-	-	-	-	-	-
Surirella	2	-	2	2	2	1	-	1	1	1	1
Actinospora (Fungi imperfecti)	-	-	-	1	-	-	-	-	-	-	-
cercariae	1	-	-	-	-	-	-	-	-	-	-
Chaoborus	-	-	-	-	-	-	-	-	-	-	1
Chironomidae	1	-	1	1	1	1	1	1	1	1	-
detritus (3=much, 4=abundant)	3	4	4	4	4	3	4	4	4	3	3
Ephemerida	1	-	-	1	1	1	1	1	-	1	-
iron bacteria	1	-	1	1	-	-	-	-	-	-	-
Odonata eggs	2	-	-	1	1	1	1	1	1	-	-
Plumatella zooecia	1	-	1	1	-	-	-	-	-	1	-
Simulidae	-	-	1	-	-	-	-	-	-	-	-
sponges spicula	1	-	1	1	1	1	3	1	-	-	-
Trichoptera	-	-	-	1	-	-	-	-	-	-	-

Table III. Plankton from the Tibiti and Coppename Rivers (25. I. 1964)

	Tibiti 1 hour upstream Sabana	Tibiti swamp	Tibiti near Sabana	Coppename near Tibiti
C r u s t a c e a				
Alona intermedia	1	-	-	-
Bosmina	1	1	-	-
Calanoid Copepods	-	-	1	1
Chydorus	-	1	-	-
Cyclops	1	-	-	-
Diaptomus	-	1	-	-
Scapholeberis mucronata	-	-	1	-
Zoea larvae	-	1	1	-
R o t a t o r i a				
Cathypna luna	-	1	-	-
Cephalodella	1	-	-	-
Filinia longiseta	1	-	-	-
Hexarthra insulina	1	1	1	-
Lecane	-	1	1	-
Lepadella cf. branchicola	1	1	1	-
Monostyla bulla	1	-	1	-
Trichocerca bicarinatus	1	-	-	-
Trichocerca indeterm.	1	-	-	-
P r o t o z o a				
Arcella	1	1	-	-
Difflugia	-	1	1	-
Euglypha	1	-	1	-
Heliozoa	1	-	-	-
C h l o r o p h y c e a				
Dictyosphaerium pulchellum	1	1	2	-
D e s m i d i a c e a e				
Closterium acicularis	2	1	2	-
Cosmarium	-	1	1	-
Micrasterias arcuatus	1	1	-	-
Micrasterias indeterm.	-	1	-	-
Pleurentaenium	1	-	-	-
D i a t o m e a e				
Actinocyclus normanni	1	1	1	1
Coscinodiscus	-	-	-	3
Diatoma	1	1	1	-
Eunotia asterionelloides	-	1	-	-
Melosira granulata	-	1	1	-
pennatae indeterm.	-	1	-	-
Pinnularia	-	1	-	-
Pleurosigma ?	-	-	-	1
Surirella	1	1	1	-
Synedra	1	1	-	-
C y a n o p h y c e a e				
Oscillatoria	1	-	-	-
Actinospora (Fungi imperfecti)	-	-	1	-
Ceratopogon	1	-	-	-
Ephemerida larvae	1	-	-	-
iron bacteria	3	1	1	-
Plumatella zooecia	1	-	-	-
sponges spicula	3	1	1	2

Table IV. Plankton from four rivulets crossing the road Afobaka-Brownsweg (26. XI. 1963)

	Rivulet A	Rivulet B	Rivulet C	Rivulet D
C r u s t a c e a				
Alona	-	-	1	-
Harpacticidae	-	1	-	-
Ostracoda	-	-	1	-
R o t a t o r i a				
Cephalodella gibba	-	1	-	'-
Dipleuchlanis propatula	-	-	1	-
Euchlanis dilatata	-	-	1	-
Lecane stenroosi	-	-	-	1
Lecane indeterm.	1	-	-	-
Lepadella patella	-	-	-	1
Monostyla quadridentata	-	-	1	-
Rotifer neptuneus	1	-	-	-
indeterm.	-	1	-	-
P r o t o z o a				
Arcella	-	1	1	1
C h l o r o p h y c e a				
Batrachospermum	-	1	1	1
Mougeotia	-	3	1	3
Spirogyra	1	-	-	1
Zygnema	-	-	1	-
D e s m i d i a c e a e				
Closterium	-	1	2	2
Euastrum	-	-	-	1
Pleurentaenium	-	1	2	3
Staurastrum	-	1	-	-
D i a t o m e a e				
Fragilaria construens	-	1	-	-
Surirella	-	1	-	1
Synedra ?	-	1	-	-
pennatae	1	1	1	-
C y a n o p h y c e a e				
Oscillatoria	-	1	-	-
Lyngbya	-	1	-	-
Rivularia ?	-	1	-	-
Actinospora (Fungi imperfecti)	-	1	1	1
Doryssa	-	1	-	-
fish	1	-	-	3
mosquito larvae	-	1	-	-
Nematoda	-	1	1	-
Odonata eggs	-	1	-	-
Oligochaeta	-	1	-	-
Plecoptera ? larvae	-	1	-	-
tadpoles	2	-	-	-

Table V. Plankton from Carolina Kreek and Compagnie Kreek (1963)

	Carolina Kreek 16. XI.	Carolina Kreek 7. XII.	Compagnie Kreek 20. XI.
C r u s t a c e a			
Alona	-	1	-
Cyclops	1	1	2
Harpacticidae	-	-	1
R o t a t o r i a			
Cathypna luna	-	1	-
Cephalodella mucronata	-	1	1
Conochiloides coenobasis	-	-	1
Dinocharis pocillum	-	-	1
Diurella	-	1	-
Filinia longiseta	-	-	2
Lecane stokesi	-	1	-
Lecane indeterm.	1	1	-
Macrochaetus	-	1	-
Monommata longiseta	-	1	-
Monostyla bulla	-	1	-
Monostyla quadridentata	-	-	1
Platyias quadricornis	1	-	-
Trichocerca	-	1	-
indeterm.	1	-	-
P r o t o z o a			
Arcella	1	1	1
Heliozoa	-	1	-
C h l o r o p h y c e a			
Characium	1	1	-
Mougeotia	-	1	-
Spirogyra	1	1	1
D i a t o m e a e			
Diatoma	-	1	-
Fragilaria capucina	-	-	2
Surirella	-	-	1
Synedra	-	1	-
D e s m i d i a c e a e			
Desmidium	-	1	1
Closterium	-	-	2
Euastrum	-	-	1
Pleurentaenium	-	-	1
Sphaerozosma granulatum	-	-	1
C y a n o p h y c e a e			
Oscillatoria	-	-	2
Actinospora (Fungi imperfecti)	-	1	1
Chlamydomonas	-	-	1
Nematoda	-	-	1
Pristina longiseta	-	1	-

Table VI. Plankton from Witte Kreek and 2nd Kreek near Kabel (18.II.1964)

	Witte Kreek	2nd Kreek shallow part	2nd Kreek deeper part, dead leaves
C r u s t a c e a			
Alona	-	-	1
Bosminopsis deitersi	3	1	1
Ceriodaphnia cornuta	1	1	-
Chydorus sphaericus	-	-	1
Diaptomus	1	4	1
Moina	1	-	1
R o t a t o r i a			
Anureopsis fissa	-	1	-
Ascomorpha saltans	-	2	-
Brachionus mirabilis	-	-	1
Filinia longiseta	-	1	-
Keratella americana	1	-	-
Lecane ludwigi	-	-	1
Lecane indeterm.	1	1	-
Lepadella patella	1	-	-
Macrochaetus	-	-	1
Monostyla bulla	-	-	1
Mytilina mucronata	-	-	1
Pedalion mira	-	-	1
Polyarthra	-	1	-
Platyias quadricornis	1	-	-
Testudinella patina	-	-	1
Testudinella brycei	-	-	1
Trichocerca similis	-	1	-
P r o t o z o a			
Arcella sp. 1	1	-	-
Arcella sp. 2	-	2	3
Difflugia	1	-	1
F l a g e l l a t a			
Eudorina elegans	-	1	1
Euglena	-	1	-
Rhipidodendron huxleyi	-	-	1
Trachelomonas caudata	-	2	-
Trachelomonas armata	-	1	-
C h l o r o p h y c e a			
Spirogyra	1	-	-
D e s m i d i a c e a e			
Closterium	1	1	2
Cosmarium	-	1	-
Docidium	-	-	1
D i a t o m e a e			
Eunotia asterionelloides	1	1	-
Surirella	-	1	-
C y a n o p h y c e a e			
Oscillatoria	1	1	-
Actinospora (Fungi imperfecti)	1	-	-
cercariae	1	-	-
Ceratopogon	1	-	-
Ephemerida	1	-	-
Leptothrix ?	1	1	-
mosquito larvae	1	-	-
Nematoda	1	-	-
sponges spicula	-	-	1
Plumatella zooecia	-	-	1

Table VII. Plankton from forest swamps, grass swamps and pools at Afobaka Plant (1963)

	forest swamp 22. XI.	grass swamp 22. XI.	grass swamp 17. XII.	swamp with stumps 22. XII.	swamp with Typha 20. XI.	oxydation pond 1. XI.
C r u s t a c e a						
Cyclops	-	1	-	1	-	-
Daphnia	-	-	-	-	-	1
Moina	-	-	-	-	1	-
R o t a t o r i a						
Anureopsis fissa	1	1	1	1	1	-
Asplanchna	-	-	-	-	-	1
Brachionus calyciflorus	-	-	-	-	1	-
Conochiloides coenobasis	-	-	-	-	1	-
Dinocharis pocillum	-	-	-	-	1	-
Eudactylota eudactylota	1	-	-	1	1	-
Filinia longiseta	-	-	-	-	1	1
Keratella americana	-	-	1	-	-	-
Lecane	-	-	-	-	1	-
Monostyla bulla	1	-	-	-	-	-
Platyias patulus	-	-	-	-	1	-
Polyarthra	-	1	2	1	2	-
Trichocerca similis	-	1	-	-	1	-
Rotifer neptuneus	-	-	-	-	1	-
Rotifer indeterm.	-	-	-	-	1	-
P r o t o z o a						
Arcella	1	-	1	-	1	-
Vorticella	-	-	-	-	-	1
F l a g e l l a t a						
Eudorina elegans	-	-	-	-	5	-
Euglena spirogyrae	-	-	1	-	-	-
Euglena indeterm.	1	2	2	-	1	1
Gymnodinium	-	1	-	-	-	-
Lepocinclis	-	-	-	-	-	3
Phacus torta	-	-	-	-	1	3
Phacus aenigmaticus	-	-	-	-	-	1
Phacus indeterm.	-	1	1	-	1	1
Synura uvella	-	1	-	-	1	-
Trachelomonas armatum	1	-	1	1	1	3
Trachelomonas volvocina	2	1	2	1	1	-
Tr. acanthophora f. speciosa	-	-	1	1	-	-
indeterm.	-	-	-	-	-	3
C h l o r o p h y c e a						
Actinastrum hantzschi	-	-	-	-	-	1
Botryococcus brauni	-	2	3	-	-	-
Coelastrum cambricum	1	3	3	1	-	-
Kirchneriella obesa	-	1	1	-	-	-
Mougeotia	3	-	-	1	-	-
Pediastrum duplex	2	4	4	1	-	-
Scenedesmus arcuata	1	-	1	-	-	-
Selenastrum gracile	1	1	1	-	-	-
Sorastrum spinulosum	-	1	-	-	-	-
Spirogyra	1	1	1	-	-	3
D e s m i d i a c e a e						
Closterium	1	1	2	-	2	-
Cosmarium	-	-	1	-	-	-
Desmidium laticeps	1	-	-	-	-	-
Euastrum	1	-	1	-	1	-
Micrasterias brasiliensis	1	3	4	1	-	-
Micrasterias arcuatus	1	-	-	-	-	-
Micrasterias truncata	1	-	-	-	-	-
Pleurentaenium	-	-	2	1	-	-
Staurastrum inaequale	-	-	-	1	-	-
Staurastrum ophiura	-	1	1	-	-	-
Staurastrum indeterm.	1	1	1	1	-	-
Sphaerozosma granulatum	-	-	-	-	1	-
D i a t o m e a e						
Pinnularia viridis	1	-	3	-	-	-
Surirella	-	1	-	-	-	-
pennatae indeterm.	2	-	2	1	1	-
C y a n o p h y c e a e						
Oscillatoria	3	1	1	-	-	3
Microcystis	-	-	-	-	-	1
Spirulina	1	1	1	1	-	-
Actinospora (Fungi imperfecti)	-	-	-	-	1	-
Gastrotricha	1	-	-	-	1	-

155

Table VIII. Plankton from forest swamps, grass swamps and pools
East of Afobaka at the 4th dam (1963)

	pool with trees 19. XI.	25. XI.	pool A near road 25. XI.	17. XII.	pool B near road 25. XI.
Crustacea					
Cyclops	-	-	1	3	1
Moina	1	3	-	-	-
Rotatoria					
Anureopsis fissa	-	-	2	1	1
Filinia longiseta	-	-	1	-	-
Hexarthra insulina	-	-	2	1	2
Lecane	2	1	-	2	1
Monostyla bulla	2	1	1	2	-
Platyias patulus	-	-	1	-	-
Polyarthra	2	1	2	2	3
Testudinella patina	-	-	-	1	-
Trichocerca	1	1	2	1	2
Protozoa					
Arcella	1	1	1	1	-
Euplotes	-	-	-	2	-
Difflugia	-	-	-	2	-
indeterm.	-	-	-	3	-
Flagellata					
Eudorina elegans	4	4	-	1	-
Euglena acus	-	4	-	-	-
Euglena polymorphum	-	3	-	-	-
Euglena indeterm.	5	-	-	3	3
Pandorina morum	1	-	-	-	-
Phacus	-	-	-	1	-
Trachelomonas	4	3	-	3	3
indeterm.	-	-	3	2	3
Chlorophycea					
Chroococcus ?	-	-	1	-	-
Crucigenia	-	-	-	-	1
Scenedesmus arcuatus	-	-	1	-	-
Scenedesmus bijuga	-	-	1	-	-
Spirogyra	-	-	1	-	-
Desmidiaceae					
Closterium	-	1	1	1	-
Cosmarium	-	1	2	-	-
Euastrum	-	-	1	-	-
Phymatodocis alternans	-	-	1	-	-
Pleurentaenium	1	1	-	2	-
Staurastrum	-	1	1	-	-
Diatomeae					
Pinnularia viridis	-	-	1	-	-
pennatae indeterm.	-	-	3	-	-
Gastrotricha	-	-	1	-	1
mosquito larvae	1	1	1	-	1

Table IX. Plankton from Afobaka, Station 1 (1963/64)

month	Nov.			Dec.				January				February				March				April				May				June				July					August				Sept.
day	12	21	27	4	11	18	27	28	14	22	31	5	12	19	26	4	11	18	25	18	15	22	29	6	13	20	27	3	10	17	24	3	9	15	22	29	5	12	19	26	29
C r u s t a c e a																																									
Alona	-	1	-	-	1	-	1	-	-	-	-	-	-	-	-	-	-	-	-	-	-	-	-	-	-	-	-	-	1	1	-	-	-	-	-	-	-	-	-	-	-
Bosmina	1	1	-	1	-	1	-	-	-	-	-	-	3	1	4	4	1	1	1	1	3	2	1	1	1	1	3	1	2	1	1	1	2	1	-	2	2	-	-	1	-
Bosminopsis deitersi	1	1	-	-	1	-	-	-	-	-	-	1	2	3	3	1	1	1	2	1	1	1	1	1	1	2	2	2	1	1	-	1	1	1	-	2	1	3	2	2	-
Ceriodaphnia cornuta	1	1	-	-	1	-	-	-	-	-	-	1	2	-	2	1	2	2	-	1	2	1	2	1	1	-	1	1	2	1	-	2	1	3	-	1	1	-	-	1	-
Chydorus globosus	-	-	-	-	-	-	-	-	-	-	-	-	-	-	-	-	-	1	-	-	-	-	-	-	-	-	1	1	1	-	-	2	1	3	-	1	-	-	-	1	-
Chydorus sphaericus	1	1	1	-	1	-	1	-	-	-	1	1	1	1	2	1	3	1	1	1	1	1	2	1	2	1	1	2	1	1	-	3	3	2	-	2	2	1	1	1	-
Cyclops	1	1	-	-	1	-	-	-	-	-	1	1	2	1	3	1	3	2	3	2	3	3	2	2	3	2	3	3	3	3	-	3	3	4	-	3	1	1	1	1	-
Daphnia	-	-	-	-	-	-	-	-	-	-	-	-	-	-	-	-	-	-	-	-	-	-	-	-	-	-	-	-	-	-	-	3	2	1	-	1	2	1	1	-	-
Diaphanosoma brachyurum	-	-	-	-	-	-	-	1	1	-	-	-	2	3	3	3	3	2	1	1	2	2	1	1	2	3	3	2	3	3	-	2	1	3	-	2	3	3	2	2	1
Diaptomus	-	-	-	-	-	-	-	-	1	1	-	-	2	3	3	1	1	2	1	1	2	3	1	1	2	3	2	1	1	2	-	1	1	2	-	1	3	1	3	2	1
Euryalona occidentalis	-	-	-	-	-	-	-	-	-	-	-	-	-	-	-	-	1	1	1	1	1	1	1	1	1	1	2	1	1	1	-	1	2	3	-	1	1	1	3	1	-
Iliocryptus	-	-	-	-	-	-	-	-	-	-	-	-	-	-	-	-	-	-	-	-	-	1	1	1	1	-	1	-	1	-	-	1	-	1	-	-	-	1	-	-	-
Moina	-	-	1	-	-	-	-	-	-	-	-	1	2	1	-	2	-	1	1	1	1	1	1	1	1	2	-	1	1	-	-	1	1	1	-	1	1	-	1	1	-
Pleuroxus striatus	-	-	-	-	-	-	-	-	-	-	-	-	-	-	-	-	-	-	-	-	1	1	1	1	1	1	2	-	-	-	-	1	-	1	-	1	1	-	-	-	-
R o t a t o r i a																																									
Anureopsis fissa	-	-	-	-	1	-	-	-	-	-	-	-	1	1	-	-	1	-	1	1	1	1	1	1	-	1	-	3	-	1	-	1	1	-	-	1	2	-	-	-	-
Ascomorpha saltans	-	-	-	-	1	-	-	1	1	-	1	1	1	-	-	1	3	1	3	1	1	1	1	4	3	2	1	1	3	1	-	1	1	-	-	1	1	2	1	3	-
Asplanchna	-	-	-	-	-	-	-	-	-	-	-	-	-	-	-	-	1	1	-	1	1	1	2	2	2	2	1	2	1	2	-	1	1	1	-	1	1	1	1	2	-
Brachionus quadridentata	-	-	1	-	2	-	-	-	-	-	-	-	-	-	-	-	1	-	-	1	1	1	1	1	1	1	1	2	1	1	-	1	1	-	-	2	1	1	-	1	-
Collotheca mucosa	-	1	1	-	-	-	-	1	-	1	-	1	-	-	-	1	-	1	-	1	1	1	1	1	1	1	1	1	1	1	-	1	-	1	-	-	-	1	-	1	-
Colurella	-	-	-	1	-	-	-	-	-	-	-	1	-	-	-	1	1	-	-	1	1	1	1	1	1	1	-	1	-	-	-	1	-	-	-	-	-	-	-	-	-
Conochiloides coenobasis	1	1	-	-	1	-	-	-	1	-	1	-	1	1	1	1	2	-	1	3	1	2	1	1	1	1	1	1	2	1	-	2	1	2	-	4	1	2	-	1	-
Conochiloides dossuaris	1	1	-	-	-	-	-	1	-	1	-	-	2	-	-	-	1	1	-	1	2	1	1	1	1	2	1	1	1	1	-	1	2	4	-	4	1	-	2	1	-
Euchlanis	1	1	-	-	1	-	-	1	-	-	-	1	-	-	-	1	1	-	1	1	1	1	1	1	1	1	1	1	1	1	-	1	2	1	-	1	2	1	-	1	-
Eudactylota eudactylota	-	-	-	-	1	-	-	-	1	-	-	-	1	-	-	-	1	-	-	-	1	1	-	1	1	1	1	1	-	-	-	-	-	-	-	-	-	-	-	-	-
Filinia longiseta	1	1	1	-	1	1	-	1	1	-	1	-	1	-	-	1	1	1	-	1	1	1	1	1	1	1	1	1	-	-	-	1	-	-	-	1	1	1	-	1	-
Habrotrocha	-	-	-	-	-	-	-	1	-	1	-	1	-	-	-	-	-	-	2	-	1	1	2	-	-	-	-	1	-	-	-	1	1	1	-	1	-	-	-	-	-
Hexarthra insulana	2	1	-	2	1	1	-	-	1	-	2	-	1	1	1	1	2	2	-	2	-	1	1	1	2	2	1	1	1	1	-	2	1	1	-	1	1	1	1	1	-
Keratella americana	1	1	1	2	1	1	-	-	1	-	1	-	1	-	-	1	1	1	-	1	1	1	1	1	2	2	1	1	2	1	-	2	1	1	-	1	1	1	1	1	-
Lecane ludwigi	-	-	-	-	-	-	-	-	-	-	-	-	-	-	-	-	1	-	-	-	1	1	1	1	1	1	-	1	1	1	-	1	1	1	-	-	-	1	-	-	-
Lecane quadridentata	1	-	-	-	1	-	-	1	-	1	-	1	-	-	-	1	-	1	-	1	-	1	1	1	1	1	1	1	1	1	-	1	2	4	-	1	-	1	-	1	-
Lecane indeterm.	-	-	-	-	-	-	-	-	-	-	-	1	-	-	-	1	1	-	-	1	1	1	-	1	1	1	1	1	1	1	-	1	2	1	-	1	2	-	1	-	-
Lepadella	-	-	-	-	-	-	-	-	1	-	-	1	-	-	-	-	-	-	-	-	-	1	-	-	-	-	-	-	-	-	-	-	-	1	-	-	-	-	-	-	-
Monostyla bulla	1	-	-	-	1	-	-	-	1	-	-	1	-	-	-	1	1	-	-	-	1	1	1	-	-	1	-	1	1	-	-	1	1	-	-	1	1	1	-	-	-
Platyias patulus	-	-	-	-	-	-	-	-	-	-	-	-	-	-	-	-	1	1	-	1	-	1	-	-	-	1	-	1	-	-	-	-	-	-	-	-	-	-	-	-	-
Platyias quadricornis	1	2	2	2	1	1	-	-	2	1	-	1	1	1	2	2	3	2	1	1	1	2	2	1	1	1	1	1	1	-	-	1	3	2	-	1	1	-	1	1	-
Polyarthra	1	2	2	1	1	-	1	2	1	-	2	1	1	1	1	2	1	2	1	2	2	1	2	1	1	1	3	1	2	-	-	1	2	3	-	1	1	1	1	1	-
Rotaria	-	-	-	-	-	-	1	-	-	-	-	1	1	-	1	1	1	1	1	1	1	1	2	-	-	-	-	1	1	1	-	1	1	1	-	-	1	-	1	1	-
Sinantherina spinosa	-	-	-	-	1	-	-	-	-	1	-	-	-	-	-	-	-	-	-	-	-	-	-	-	-	-	-	-	-	-	-	-	-	-	-	-	1	-	-	-	-
Synchaeta	2	-	-	1	-	1	-	-	1	-	-	1	-	-	-	-	-	1	-	-	-	-	-	-	-	-	-	-	-	-	-	-	1	-	-	1	1	1	1	1	-
Testudinella patina	-	-	-	-	-	-	-	-	-	-	-	-	-	-	-	-	-	-	-	-	-	-	-	-	-	-	-	-	-	1	-	-	-	1	-	1	-	1	-	1	-
Trichocerca bicristata	-	1	-	1	1	1	-	-	1	1	-	-	-	-	-	-	-	1	-	-	-	-	-	-	-	-	-	-	-	1	-	-	1	-	-	1	-	1	-	1	-
Trichocerca similis	-	1	-	-	1	1	1	-	1	-	1	-	1	-	-	-	1	1	1	-	1	1	2	-	-	-	1	-	1	1	-	1	1	1	-	-	1	1	1	1	-

Table IX. (1st. continued)

Trichocerca indeterm.
Trichosphaera aequatorialis
Protozoa
Amoeba
Arcella
Centropyxis
Difflugia
Euglypha
Heliozoa
Tintinnidium
Tokophrya?
indeterm.
Flagellata
Dinobryon divergens
Eudorina elegans
Euglena
Gonium pectorale
Mallomonas
Pandorina morum
Peridinium
Phacus
Pyrobotrys
Strombomonas ensifera
Synura uvella
Trachelomonas caudata
Trachelomonas indeterm.
indeterm.
Uroglena volvox
Chlorophycea
Acanthosphaera zacharasiae
Ankistrodesmus falcatus
Chodatella
Chroococcus
Coelastrum cambricum
Crucigenia
Dictyosphaerium ehrenbergi
Dimorphococcus
Elakathotrix?
Franceia?
Golenkinia radiata
Hormidium?
Kirchneriella obesa
Mougeotia
Scenedesmus bijuga
Scenedesmus indeterm.
Spirogyra
Tetrallantos Lagerheimii
indeterm. green colonies
indeterm. filaments

158

Table IX. (2d. continued)

Desmidiaceae																							
Closteriopsis																							
Closterium indeterm.																							
Closterium kuetzingii			1							1													
Cosmarium	1	1		1	1	1	1	1	1	1	1	1	1	1	1	1	1	1	1	1			
Cosmocladium			1		1				1			1											
Desmidium swartzi			1																				
Docidium baculum					1				1	2	1	1	2	2			1	2		1			
Euastrum																							
Gonatozygon	1	1		1				1					1	2	1	1							
Hyalotheca rmucosa	1		1	1			1				1			1									
Micrasterias abrupta																							
Micrasterias arcuatus									1						1								
Micrasterias brasiliensis			1					1							1	1							
Micrasterias mahabhaalareswensis																							
Micrasterias indeterm.	1		1	1												1							
Penium	1	1				1	2				3	1	1	1	1	1	1	2	1	1	1	1	1
Diatomeae																							
Sphaerozosma granulatum	2	3	2	2	1	1							1	2	1	1	1	2	1	1	1	1	1
Staurastrum inaequale	1	3	2	2	1	1							2	1	2	3	1		1	1	1		1
Staurastrum mammillatum	2	1	2	1		1				2			1	2	3	4	3		1		1	1	
Staurastrum trifidum	1	1		1	1								1	1	1	1	1			1		1	1
Staurastrum i ndeterm.			1	1	1	2							1	1	2	2	1	1	1	1		1	1
Xanthidium antilopeum		1		1						1											1	1	1
Xanthidium fragile	1		1	1																			
Diatomeae																							
Cyclotella			1						1				2										
Diatoma	3	4	4	5	3	3	3	2	4	2	1	1		3	1	1		1	1	1	1		2
Eunotia asterionelloides	4	4	5	3	3	2	4	2	2	1					1	1	1	3	1	1	1	1	
Melosira granulata	1		3	1	1	1	1	1	1	1			3	1	1		2	1	1	1	1	2	1
Pinnularia			1	1	1	1			1	1													1
Rhizosolenia	2	1	1	1	1	1	1	3	1	1					1	1	1		3		1		
Surirella	1		1	1				1	1				2	1	2		1		1	1		1	
Cyanophyceae																							
Lyngbya							1																
Merismopedia			1	1		1	1								1		1						
Oscillatoria			1			1				1		1					1		1				
Rivularia						1	1						1				1						
Actinospora (Fungi imperfecti)																							
Aulophorus furcatus							1						1		1			1					
Catenula			1	1																			
Chaetogaster			1			1			1														
Chaoboris			1	1					1				1		1								
Ephemerida				1	1		1																
Gastrotricha			1	1					1						1								
mosquito larvae		1			1	1			1					2				1					
Nematoda			1			1	1		1				1										
Odonata eggs						1																	
Plumatella zooecia	1	1	1	1	1	1	1																
sponges spicula	1	1	1	1	1	1	1	1	1														
veliger	1	1	1	1				1	1														

Table X. Plankton from Kabel, Station 2 (1964)

month	January	February			March				April					May				June				July					August				September	
day	30	12	19	26	4	11	18	25	1	8	15	22	29	6	13	20	27	3	10	17	24	1	9	15	22	29	5	12	19	26	2	9

Crustacea
- Alonella
- Bosmina
- Bosminopsis
- Ceriodaphnia cornuta
- Chydorus
- Cyclops
- Diaphanosoma brachyurum
- Diaptomus
- Euryalona occidentalis
- Iliocryptus
- Moina
- Moinadaphnia

Rotatoria
- Anureopsis fissa
- Asplanchna
- Brachionus mirabilis
- Collotheca mutabilis
- Conochiloides coenobasis
- Euchlanis
- Eudactylota
- Filinia longiseta
- Hexarthra
- Keratella americana
- Lecane
- Lepadella
- Monommata longiseta
- Platyias patulus macrura
- Platyias quadricornis
- Polyarthra
- Rotaria
- Sinantherina spinosa
- Testudinella mucronata
- Testudinella patina
- Testudinella tridentata
- Trichocerca sp. 1
- Trichocerca sp. 2

Protozoa
- Arcella
- Centropyxis
- Difflugia
- Euglypha
- Heleozoa
- Myriophrys paradoxa

Table X. (1st continued)

Flagellata																		
Dinobryon	1	1	1	1	1	1	1	1	1	1	1	1	1	1	1	2	2	1
Eudorina elegans	1	2	1	1	1	3	4	3	2	2	3	3	1	2	2	1	2	1
Euglena acus	–	–	–	–	2	4	4	1	1	2	–	–	–	–	1	1	2	1
Euglena indeterm.	–	–	–	–	–	–	–	–	–	–	–	–	–	–	–	–	–	–
Gonium pectorale	–	–	1	–	1	2	1	–	2	1	1	2	–	–	–	–	–	–
Pandorina morum	–	–	–	–	–	1	1	–	1	1	1	–	–	1	–	1	1	1
Peridinium	–	–	2	–	1	1	1	1	1	1	1	1	1	–	1	1	1	1
Rhipidodendron huxleyi	1	–	–	–	–	1	3	1	–	3	3	–	1	3	1	1	1	1
Strombomonas ensifera	–	–	–	–	–	1	1	–	1	1	–	–	1	–	–	–	–	–
Synura	1	–	–	–	1	2	1	2	1	2	2	–	1	2	–	1	1	1
Trachelomonas caudata	–	–	–	–	1	1	1	1	1	1	1	1	1	–	–	–	–	–
Trachelomonas indeterm.	–	–	1	1	1	2	2	2	1	1	2	2	1	1	1	1	1	1
Uroglena americana	–	–	–	–	1	1	2	1	1	1	1	1	1	–	1	1	1	1
indeterm.	1	1	1	–	2	2	1	1	1	2	1	2	2	–	2	2	2	3
Chlorophyceae																		
Coelastrum microporum	1	1	1	2	1	3	3	1	1	1	1	1	–	–	1	1	1	3
Dictyosphaerium pulchellum	1	2	2	1	2	1	2	2	2	2	2	–	1	2	–	2	1	3
Dimorphococcus lunatus	1	–	–	1	1	2	1	1	1	1	1	–	1	1	–	1	1	1
Elakathothrix	–	–	–	–	1	1	1	1	1	1	1	–	1	1	–	–	–	–
Franceia?	1	1	2	2	3	2	3	1	1	1	1	–	1	2	1	1	1	1
Kirchneriella obesa	1	1	1	1	1	1	1	1	1	1	1	1	1	3	1	1	1	1
Micractinium pusillum	–	1	1	1	1	1	1	1	1	1	1	–	1	–	1	1	1	1
Microspora	1	–	–	–	–	1	1	1	1	1	–	–	1	–	–	–	–	1
Mougeotia	–	1	–	1	1	1	1	1	1	1	1	1	1	1	–	–	1	–
Scenedesmus bijuga	1	1	1	1	1	1	1	1	1	1	1	1	1	1	–	1	–	1
Scenedesmus indeterm.	1	1	1	1	1	1	1	1	1	1	1	–	1	1	–	1	–	1
Selenastrum gracile	1	1	1	–	1	2	1	1	1	1	1	1	1	1	–	1	–	1
Spirogyra	1	1	–	–	1	–	–	–	1	–	–	–	1	–	–	–	–	2
green colonies	1	–	–	–	–	1	–	1	1	1	1	–	1	–	–	–	–	3
Desmidiaceae																		
Closterium kuetzingii	1	–	1	1	1	1	1	1	1	1	1	1	1	1	–	–	–	–
Closterium indeterm.	–	1	1	1	1	1	1	1	1	1	1	1	1	1	–	–	–	–
Cosmarium	–	1	1	1	1	1	1	1	1	1	1	1	1	1	–	–	2	1
Cosmocladium	1	1	2	3	1	1	2	1	2	1	2	1	1	2	–	–	2	1
Euastrum	1	1	1	1	1	1	1	1	1	1	1	1	1	1	–	–	1	1
Gonatozygon	1	1	1	1	1	1	1	1	1	1	1	1	1	1	–	–	1	1
Hyalotheca	1	–	1	2	3	2	2	1	1	1	1	1	1	1	–	–	2	1
Micrasterias arcuatus	1	1	1	1	1	1	3	1	1	1	1	1	1	1	–	–	1	1
Schizacanthum?	2	2	2	1	1	2	2	1	1	1	1	1	1	1	1	1	1	1
Sphaerozosma	1	1	3	1	1	3	3	1	1	1	1	1	1	1	–	–	1	1
Staurastrum mamillatum	2	1	2	1	1	2	1	1	1	1	1	1	1	1	–	–	1	1
Staurastrum trifidum	1	2	1	1	1	1	1	1	1	1	1	1	1	1	–	–	1	1
Staurastrum inaequale	2	1	2	2	2	2	1	1	1	1	1	1	1	1	–	–	1	1
Staurastrum indeterm.	1	1	1	1	2	2	1	2	1	1	1	1	1	2	–	–	1	1
Diatomeae																		
Actinocyclus normanni	–	–	–	–	–	–	–	–	1	1	1	1	–	–	–	–	–	–
Cyclotella	1	–	–	–	–	–	–	–	–	–	–	–	1	–	1	–	–	1
Diatoma	1	–	–	–	–	–	–	–	–	–	–	–	1	1	–	–	–	–

Table X. (2nd. continued)

Eunotia asterionelloides	3	4	4	3	3	2	2	2	3	4	3	3	2	2	1	1	1	1	1	2	3	4	3	4	1	2	1	-
Melosira granulata	-	-	1	1	-	2	1	1	1	1	1	1	1	-	1	1	-	2	-	1	3	3	3	4	3	4	4	3
Rhizosolenia	-	2	1	2	2	1	2	1	2	2	1	1	1	1	1	-	2	1	1	3	2	3	2	3	-	-	-	-
Surirella	1	1	1	1	-	1	1	-	2	-	1	-	1	1	1	-	-	-	3	-	-	-	-	-	-	-	-	-
Cyanophyceae
Gloeocystis	-	1	-	-	-	-	-	1	-	-	-	-	-	-	-	-	-	-	-	1	-	-	-	-	-	1	-
Merismopedia convoluta	1	1	-	-	-	1	-	-	-	-	-	-	-	-	-	-	-	-	-	-	-	-	-	-	-	-	-
Nodularia?	1	-	-	-	-	-	-	-	-	-	-	-	-	-	-	-	-	-	-	-	-	-	-	-	-	-
Oscillatoria	1	-	-	-	-	-	-	-	-	1	-	-	-	1	-	-	-	-	-	-	-	-	-	-	-	-
Chaoborus	-	-	-	-	-	1	1	-	-	-	-	-	-	1	-	-	-	-	-	-	-	-	-	-	-	-
Ephemerida	1	1	1	1	1	1	1	-	-	-	-	-	-	-	-	-	-	-	-	-	-	-	-	-	-	-
mosquito larvae	1	-	-	1	1	-	-	-	-	-	-	-	-	-	-	-	-	-	-	-	-	-	-	-	-	-
Odonata eggs	1	-	-	-	-	-	-	-	-	-	-	-	-	-	-	-	-	-	-	-	-	-	-	-	-	-
Podostemaceae (pieces)	1	1	1	-	-	-	-	1	1	1	-	-	-	1	-	-	-	-	-	-	-	-	-	-	-	-
sponges spicula	1	-	1	1	1	-	-	1	-	-	-	-	-	-	-	-	-	-	-	-	-	-	-	-	-	-
veliger	-	-	-	1	1	-	1	-	-	-	-	-	-	-	-	-	-	-	-	-	-	-	-	-	-	-

Table XI. Plankton from Beerdotti, Station 5 (1964)

month	April	June		July					August				September	
day	30	17	24	1	9	15	22	29	5	12	19	26	2	9
Crustacea														
Alona	-	1	-	-	-	-	-	1	-	-	-	-	-	-
Alonella dadayi	-	-	-	-	-	-	-	1	-	-	-	-	-	-
Alonella indeterm.	-	-	-	-	-	-	-	1	-	-	-	-	-	-
Bosminopsis deitersi	-	-	-	-	-	-	1	1	1	1	1	-	-	-
Cyclops	1	-	-	-	-	-	1	1	1	1	2	2	1	1
Diaphanosoma brachyurum	-	-	-	-	-	-	1	1	1	1	2	2	-	1
Diaptomus	-	-	-	-	-	-	-	-	-	-	-	1	1	2
Graptoleberis testudinaria	-	-	-	-	-	-	-	1	-	-	-	-	-	-
Harpacticidae	-	-	-	-	-	-	-	1	-	-	-	-	-	-
Iliocryptus	-	-	-	-	-	-	-	-	-	1	-	-	-	-
Moina	-	-	1	1	-	-	1	1	-	1	1	-	-	1
Rotatoria														
Anureopsis fissa	-	-	-	-	-	-	-	-	-	1	-	-	-	-
Asplanchna	-	-	-	-	-	-	-	-	-	-	-	-	-	1
Conochiloides coenobasis	-	-	-	-	1	-	-	-	-	1	1	2	1	1
Euchlanis	-	-	-	-	1	-	-	1	1	-	-	-	-	-
Eudactylota eudactylota	-	-	-	-	-	-	-	2	-	-	-	-	-	-
Filinia longiseta	-	-	-	1	1	-	1	1	2	2	2	1	1	1
Hexarthra	-	-	1	-	-	-	1	1	1	1	1	1	-	1
Keratella americana	-	-	-	-	-	-	-	-	-	1	-	-	-	-
Lecane	-	-	-	-	-	-	1	1	-	-	-	-	-	-
Platyias quadricornis	-	-	-	-	-	-	-	1	-	-	-	-	-	-
Polyarthra	1	-	1	1	1	-	-	1	1	1	-	1	1	2
Rotaria	-	-	-	-	-	-	-	1	-	-	-	-	-	-
Testudinella brycei	-	-	-	-	-	-	-	1	-	-	-	-	-	-
Trichocerca	-	-	-	1	-	-	-	-	-	-	-	-	-	-
Protozoa														
Arcella	-	1	-	-	1	-	1	-	-	1	-	-	-	1
Difflugia	1	-	-	-	-	-	-	-	-	-	-	-	-	-
Euglypha	-	1	-	1	-	-	-	-	-	-	-	-	-	-
Heleozoa	2	1	-	-	-	-	-	-	-	1	-	-	-	-
Flagellata														
Dinobryon	-	-	-	-	-	-	-	-	-	-	-	-	1	-
Eudorina elegans	1	1	1	1	1	-	1	1	1	1	1	1	1	1
Euglena acus	-	-	-	-	-	-	-	-	-	-	-	1	-	-
Peridinium	1	-	-	-	-	-	-	-	-	-	-	-	-	-
Rhipidodendron huxleyi	-	2	-	-	-	-	-	-	-	-	-	-	-	-
Strombomonas ensifera	-	-	1	-	-	-	-	-	-	-	-	-	-	-
Trachelomonas	-	-	-	-	-	-	-	-	-	-	-	-	-	1
Chlorophycea														
Dictyosphaerium pulchellum	-	1	-	1	-	-	-	-	-	-	-	1	-	2
Mougeotia	1	-	-	-	-	-	-	-	-	-	-	-	-	-
Scenedesmus bijuga	-	-	-	-	-	-	-	-	-	-	-	-	1	-
Scenedesmus indeterm.	-	-	-	-	-	-	-	-	-	-	-	1	-	-
Desmidiaceae														
Closterium setaceum	1	-	-	-	-	-	-	-	-	-	-	-	-	-
Closterium indeterm.	-	1	1	-	-	-	1	-	-	1	-	1	1	-
Cosmarium	1	-	-	-	-	-	-	-	-	-	-	-	-	-
Euastrum	-	-	-	-	-	-	-	-	-	1	-	-	-	-
Micrasterias arcuatus	-	-	-	-	-	-	-	-	-	-	-	-	-	1
Schizacanthum	1	-	-	-	-	-	-	-	-	-	-	-	-	-
Sphaerozosma	1	-	-	-	-	-	-	-	-	-	-	-	-	-
Staurastrum mamillosum	1	-	1	-	-	-	-	-	-	-	-	-	-	-
Staurastrum trifidum	1	-	-	-	-	-	-	-	-	-	-	-	-	-
Staurastrum indeterm.	-	-	-	-	-	-	-	-	-	-	-	-	-	1
Diatomeae														
Actinocyclus	-	-	-	1	-	-	-	-	-	-	1	-	-	-
Coscinodiscus	-	-	-	-	-	-	-	-	-	-	1	-	-	-
Diatoma	-	-	-	-	-	-	-	-	1	1	-	-	-	-
Eunotia asterionelloides	4	-	1	-	1	-	-	1	-	-	1	1	-	1
Melosira	-	-	1	-	-	-	-	-	-	1	1	1	1	3
Rhizosolenia	1	-	-	-	-	-	-	-	-	-	1	1	-	2
Stephanodiscus	-	-	-	-	-	-	-	-	-	-	-	-	1	2
Surirella	1	-	-	1	-	-	-	-	-	-	-	-	-	-
Chaoborus	-	-	-	-	-	-	1	-	1	-	-	-	-	-
mosquito larvae	-	1	-	-	-	-	1	1	-	-	-	-	-	-
sponges spicula	1	-	-	1	-	-	-	-	-	-	-	1	-	-
veliger	1	-	-	-	-	-	-	-	-	-	-	-	-	-

Table XII. Plankton from Locus, Station 3 (1964)

month	February				March				April					May				June				July					August				September	
day	8	12	19	26	4	11	18	25	1	8	15	22	29	6	13	20	27	3	10	17	24	1	9	15	22	29	5	12	19	26	2	9
C r u s t a c e a																																
Bosminopsis deitersi	-	1	2	-	-	-	-	-	-	-	-	-	-	-	-	-	-	-	-	-	-	-	-	-	-	-	-	-	-	-	-	-
Ceriodaphnia cornuta	-	-	-	-	-	-	-	-	-	-	-	-	-	-	-	-	-	-	-	-	-	-	-	-	-	-	-	-	-	-	-	-
Chydorus	-	-	-	-	-	-	-	-	-	-	-	-	1	-	-	-	-	-	-	-	-	-	-	-	-	-	-	-	-	-	-	-
Cyclops	-	1	1	1	1	1	1	1	1	1	1	1	1	-	-	1	-	1	1	1	1	1	2	2	3	1	3	3	3	3	3	3
Diaptomus	-	2	1	1	1	2	-	1	1	1	1	1	1	1	1	1	-	1	1	1	2	1	1	1	1	1	3	1	1	2	2	3
Diaphanosoma brachyurum	-	1	1	2	2	-	1	1	1	1	1	1	1	1	1	1	-	1	1	1	1	1	1	1	1	1	1	1	2	3	2	3
Moina	-	-	1	-	-	-	1	1	-	-	-	-	-	-	-	-	-	-	-	-	-	-	-	-	-	-	-	-	-	1	-	1
Ostracoda	-	-	-	-	-	-	-	-	-	-	-	-	-	-	-	-	-	-	-	-	-	-	-	-	-	-	-	-	-	-	-	-
R o t a t o r i a																																
Anureopsis fissa	1	-	1	-	-	-	-	-	1	-	1	1	1	-	-	-	-	-	-	-	-	-	-	-	-	-	-	1	1	1	1	1
Ascomorpha saltans	-	1	1	2	-	1	1	-	-	-	-	-	-	-	-	-	-	-	-	-	-	-	-	-	-	-	-	-	-	-	1	1
Asplanchna	1	1	1	1	2	1	1	1	1	1	1	1	1	1	1	1	1	1	1	1	1	1	1	1	1	1	1	1	1	1	1	1
Brachionus falcatus	-	-	-	-	-	-	-	-	-	-	-	-	-	-	-	-	-	-	-	-	-	-	-	-	1	-	-	2	-	2	-	-
Brachionus mirabilis	-	-	-	-	-	-	-	-	-	-	-	-	-	-	-	-	-	-	-	-	-	-	-	-	-	-	-	-	-	-	-	-
Brachionus quadridentata	-	1	-	-	-	-	-	-	-	-	-	-	1	1	1	-	-	2	-	-	-	1	-	-	-	-	1	1	1	1	1	1
Collotheca mutabilis	-	-	-	-	-	-	-	-	-	-	-	-	-	-	-	-	-	-	-	-	-	-	-	-	-	-	-	-	-	-	-	-
Colurella	-	1	1	-	-	-	-	1	1	1	1	1	1	1	1	1	1	1	1	1	1	1	1	1	1	1	1	1	1	1	1	1
Conochiloides coenobasis	-	1	1	1	-	1	1	1	1	1	1	1	-	-	-	-	-	1	1	1	1	1	1	1	1	2	2	2	2	1	1	1
Euchlanis	-	1	2	2	3	2	1	-	2	2	1	1	1	-	-	-	-	-	-	-	-	-	-	-	1	-	2	2	2	1	1	1
Filinia longiseta	-	2	1	-	-	1	-	-	-	1	1	1	1	-	1	-	-	-	-	-	1	-	-	-	-	-	-	-	-	-	-	-
Hexarthra	2	1	-	-	-	-	-	-	-	-	-	-	-	-	-	-	-	-	-	-	-	-	-	-	-	-	-	-	-	-	-	-
Lecane ludwigi	-	1	1	-	-	1	-	1	1	1	1	1	1	1	1	1	1	1	1	1	1	1	1	1	1	-	1	-	-	1	1	1
Lecane quadridentata	1	1	1	-	2	-	-	-	1	-	-	-	-	2	-	1	-	2	-	-	-	-	-	-	-	-	1	-	-	-	-	-
Lecane indeterm.	1	1	1	-	-	-	1	1	1	1	1	1	1	1	1	1	1	1	1	1	1	1	1	1	1	1	1	1	1	1	1	1
Lepadella	1	1	-	-	-	-	-	1	1	-	-	-	-	-	-	-	-	-	-	-	-	-	-	-	-	-	1	-	-	-	-	-
Monommata	-	-	-	2	2	1	1	1	1	1	1	1	1	1	1	1	1	1	1	1	1	1	1	1	1	2	1	1	1	1	1	1
Monostyla bulla	-	-	-	-	-	-	-	-	-	-	-	-	-	-	-	-	-	-	-	-	-	1	-	-	-	-	-	1	-	1	-	-
Mytilina	-	1	-	-	-	1	1	1	1	-	-	-	-	-	-	-	-	-	-	-	-	1	-	-	-	-	1	-	-	-	-	-
Platyias patulus	1	1	1	2	1	1	-	1	1	1	-	-	-	1	1	1	2	-	1	1	1	-	1	1	1	1	1	-	1	-	1	1
Platyias quadricornis	-	2	1	1	2	1	1	-	-	-	-	-	-	-	-	-	-	-	-	-	-	1	2	1	1	1	-	1	1	1	1	-
Polyarthra	-	-	-	-	2	1	1	-	1	-	1	-	-	-	-	-	-	-	-	-	1	1	1	1	3	-	1	-	-	-	-	-
Rotaria	-	1	1	1	-	1	1	1	1	1	1	1	1	1	1	1	1	1	1	1	1	1	1	1	1	1	1	1	1	1	1	1
Sinantherina spinosa	-	1	1	1	1	-	1	-	1	1	1	-	-	-	-	-	-	-	-	-	-	1	-	1	-	-	1	1	-	1	-	-
Synchaeta	-	-	-	-	1	1	1	1	1	1	1	-	-	-	-	-	-	-	-	-	-	-	-	-	-	-	-	-	-	-	-	-
Trichocerca	-	1	1	-	-	1	1	1	1	1	1	1	1	-	1	1	1	-	1	1	1	1	1	1	1	-	1	-	-	1	1	-
Trochosphaera	-	-	-	-	-	-	-	-	-	-	-	-	-	-	-	-	-	-	-	-	-	-	-	-	-	-	-	-	-	1	1	-
indeterm.	-	-	-	-	1	1	,	-	1	1	1	-	-	-	-	-	-	-	-	-	-	-	-	-	-	-	-	-	-	-	-	-
P r o t o z o a																																
Arcella	1	1	2	3	2	1	-	-	2	2	1	1	1	1	-	-	-	-	-	-	-	-	1	1	-	1	1	1	1	1	1	1
Heleozoa	1	1	1	1	1	-	-	-	2	1	-	-	-	-	-	-	-	1	-	-	-	1	1	1	-	-	-	-	-	-	-	-
Pelomyxa	-	-	-	-	-	-	-	-	-	-	-	-	-	-	-	-	-	-	-	-	-	1	-	-	-	-	-	-	-	-	-	-
Tokophrya ?	-	1	1	1	1	1	-	1	1	1	-	1	1	1	1	2	-	1	1	-	-	-	-	1	-	-	-	-	-	-	-	-
Vorticella	-	1	1	1	1	-	1	1	1	1	-	1	-	-	1	2	-	-	-	-	1	-	-	-	-	1	-	2	-	-	-	-
indeterm.	-	-	-	-	-	-	-	-	-	-	-	-	-	-	-	-	-	-	-	-	-	-	-	-	-	-	-	-	-	-	-	-

163

Table XII. (1st. continued)

Flagellata

	1	2	3	4	5	6	7	8	9	10	11	12	13	14	15	16	17	18	19	20	21	22	23	24	25	26	27	28	29	30	31	32	33	34	35	36	37	38	39	40
Eudorina elegans	1	1	3	3	1	1	1	1	1	1	1	1	1	1	-	-	1	1	3	3	3	1	1	-	-	1	-	-	-	1	-	-	-
Euglena acus ·	-	-	1	1	1	1	1	-	-	-	-	-	-	-	-	-	-	-	-	-	-	-	-	-	-	-	-	-	-	-	-	-
Euglena oxyuris	-	-	1	-	-	-	1	-	-	-	-	-	-	-	-	1	-	-	-	-	-	1	-	-	-	-	-
Euglena indeterm.	-	-	1	-	-	-	-	1	1	3	3	1	-	1	2	-	1	4	1	-	-	-	-	-	-	-	-
Mallomonas	-	-	-	-	-	-	-	-	-	-	-	-	-	-	1	-	-	-	-	-	-	-	1	-	3
Pandorina morum	-	-	1	1	-	-	-	-	-	-	-	-	-	-	-	-	-	-	-	-	-	-	-	-
Peridinium	-	-	-	-	-	-	-	-	-	-	1	-	-	-	-	-	-	-	-	-	-	-	-
Phacus longicauda	-	-	1	-	-	-	1	-	-	-	1	-	-	-	-	-	1	-	-	-	-	-	-
Phacus pleuronectes	-	-	-	-	1	1	-	1	-	1	1	-	-	-	-	-	-	-	-	-	-	-
Phacus torta	-	-	1	-	-	1	-	-	1	-	1	-	1	-	-	-	1	-	-	-	-	-
Pyrobotrys	-	-	-	-	-	-	1	-	1	-	2	1	1	1	1	1	-	-	-	-	-
Strombomonas ensifera	-	-	-	-	1	-	1	1	1	2	2	2	2	2	1	3	2	2	2	-	1	-	-	1	-	1
Trachelomonas caudata	1	1	1	-	1	-	-	1	-	-	-	-	-	-	-	-	-	-	-	-	-	-
Trachelomonas indeterm.	-	-	-	1	-	1	-	1	2	1	2	3	3	3	2	2	1	1	2	2	2	1	2	1	1	1	-	1	2	2
Uroglena	-	-	1	-	-	-	-	-	-	-	-	-	-	-	-	-	-	-	-	-
indeterm.	-	-	3	3	-	2	1	-	-	2	1	-	2	2	2	-	2	2	2	-	1	1	-	3	-	3	-	-	-

Chlorophycea

Ankistrodesmus	-	-	-	-	-	-	-	-	-	-	-	-	-	-	-	-	1	-	-	1	-	-	-	-	-	-
Dictyosphaerium	-	-	1	-	-	-	-	-	1	1	1	2	-	1	2	2	-	2	1	1	2	2	1	1	1	-	1	1	2	2	2
Dimorphococcus	-	-	-	-	-	-	1	-	-	-	-	-	-	-	1	-	1	1	1	-	-	-	-	-
Kirchneriella obesa	-	-	-	-	-	-	-	1	-	-	-	1	-	-	-	-	1	-	1	-	-	1
Mougeotia	-	-	-	-	-	-	1	-	2	2	-	-	-	1	1	1	2	2	-	1	1	-	-	-	-	1
Scenedesmus	-	-	-	-	-	-	-	-	-	-	-	-	-	-	-	-	-	1	-	-	-	-
Spirogyra	2	-	-	1	-	1	1	-	1	1	-	1	1	-	-	1	-	1	-	-	1	2	-	-	-	-	-
Tetrallantos	-	-	-	-	-	-	-	-	-	-	-	-	-	-	-	-	-	-	-	-	-	1	1	-
indeterm. colonies	-	-	-	-	-	-	-	-	-	-	-	-	-	-	-	-	3	3	2	3	3	1	3
indeterm. filaments	-	-	-	1	-	-	1	-	1	-	-	-	-	-	-	-	-	-	-	-	1	-	-

Desmidiaceae

Closterium kuetzingii	-	-	-	-	1	-	-	1	-	1	1	-	-	-	-	-	-	-	1	-	-	-	-
Closterium indeterm.	1	1	1	1	1	1	2	2	3	3	3	3	2	2	2	2	2	3	3	2	3	1	1	-	1	1	1	1	1	1	1
Cosmarium	1	-	-	-	-	-	-	-	-	-	-	-	-	-	-	-	-	-	-	-	-	-
Cosmocladium	-	-	-	-	-	-	-	-	-	-	1	1	1	-	-	-	1	-	1	-	-	-	-
Gonatozygon	-	-	1	-	-	-	-	1	-	-	1	-	-	-	-	-	1	-	-	-	-	-
Hyalotheca	-	-	-	-	-	-	-	-	-	-	-	-	1	-	-	-	1	-	-	-	-	-
Micrasterias brasiliensis	-	-	-	-	-	-	-	-	-	-	-	-	-	-	-	1	-	-	-	-	-
Sphaerozosma	-	-	-	-	-	-	-	-	-	-	-	-	-	-	-	1	-	1	1	1	-	1
Staurastrum trifidum	-	-	-	-	-	-	-	-	-	-	-	-	-	-	-	1	-	1	1	1	1	2
Staurastrum indeterm.	-	-	-	-	-	1	-	-	-	-	-	-	-	-	1	-	1	-	1	1	1	2	1	2
Xanthidium antilopeum	-	-	-	-	-	-	-	-	-	-	-	-	-	-	-	-	-	-	-	-	-	1

Diatomeae

Cyclotella	-	-	-	-	-	-	-	-	-	-	-	-	-	-	-	2	-	-	-	-	-
Eunotia asterionelloides	1	1	-	-	-	-	1	-	-	-	1	-	-	-	-	-	1	1	-	1	-	1	-	-	-	-
Melosira granulata	-	-	1	1	-	-	-	-	-	-	-	-	-	1	-	-	-	2	2	2	1	1	1	1
Surirella	1	-	-	-	-	-	-	-	-	-	-	-	-	-	-	-	-	-	-	-	-
Synedra	-	1	-	-	-	-	-	-	-	-	-	-	-	-	-	-	-	-	-	-	-
indeterm.	-	-	-	-	-	-	-	1	-	-	-	-	-	-	-	-	-	-	-	-	-

164

Table XII. (2nd. continued)

Cyanophyceae																																
Anabaena	-	-	-	-	-	-	-	-	-	-	-	-	-	-	-	-	-	-	-	-	-	-	-	-	-	-	-	-	-	1	-	-
Oscillatoria	1	-	-	-	-	-	-	1	-	-	-	-	-	-	-	-	-	-	-	-	-	-	-	-	-	-	-	-	-	-	-	-
indeterm. filaments	-	-	-	-	-	1	1	2	2	3	1	1	1	1	1	1	1	1	3	2	2	3	2	2	-	-	-	-	-	-	-	-
Aulophorus	-	-	-	-	-	-	-	-	-	-	-	-	-	-	-	-	-	-	1	-	-	-	1	-	-	-	-	-	-	-	-	-
Catenula	-	-	-	-	-	-	-	-	-	-	-	-	-	-	-	-	-	-	-	-	-	-	-	1	1	-	-	-	4	-	-	1
Chaoborus	-	-	-	-	-	1	-	-	-	-	-	-	-	-	-	-	-	-	-	-	-	-	-	-	-	-	-	-	-	-	-	-
Gastrotricha	-	-	-	-	-	-	-	-	-	-	-	-	-	1	-	-	-	-	-	-	-	1	-	-	-	-	-	-	-	-	-	-
Lepthotrix	1	-	3	3	4	1	1	-	3	1	1	2	-	-	-	2	-	-	-	-	-	-	-	-	-	-	-	-	-	-	-	-
mosquito larvae	-	-	-	-	1	-	-	-	-	-	-	-	-	-	-	-	-	-	-	-	-	-	-	-	-	-	-	-	-	-	-	-
Nematoda	1	-	-	-	-	-	-	-	-	-	-	-	-	-	-	-	-	-	-	-	-	-	-	-	-	-	-	-	-	-	-	-
Turbellaria	-	-	-	-	3	-	-	-	-	-	-	-	-	-	-	-	-	-	-	-	-	-	-	-	-	-	-	-	-	-	-	-
veliger	-	1	-	-	-	-	-	-	-	-	-	-	-	-	-	-	-	-	-	-	-	-	-	-	-	-	-	-	-	-	-	-

Table XIII. Plankton from Station Sara (1963/64)

	Nov.		December				January					February				March				April					May				June				July					August				Septemb.	
day	21	27	4	11	18	27	2	8	14	22	31	5	12	19	26	4	11	18	25	1	8	15	22	29	6	13	20	27	3	10	17	24	1	9	15	22	29	5	12	19	26	2	9
C r u s t a c e a																																											
Bosmina	-	-	-	-	-	-	-	-	-	-	1	-	2	2	-	-	-	-	-	-	-	-	-	-	-	-	-	-	-	-	-	-	-	-	-	-	-	-	-	-	-	2	2
Bosminopsis deitersi	-	-	-	-	-	-	-	-	-	-	-	-	-	-	-	-	-	-	-	-	-	-	-	-	-	-	-	-	-	-	-	-	-	-	-	-	-	-	-	-	-	-	-
Ceriodaphnia cornuta	-	-	-	-	-	-	-	-	-	-	-	-	-	-	-	-	-	-	-	-	-	-	-	-	-	-	-	-	-	1	1	1	1	1	1	-	-	-	1	1	1	-	-
Chydorus	-	-	1	1	-	1	-	-	1	-	-	-	2	-	-	-	1	-	1	1	1	1	1	-	2	1	-	1	1	1	1	1	3	2	2	4	3	1	1	1	1	2	1
Cyclops	1	1	1	1	1	-	-	1	1	1	-	-	1	1	1	1	1	1	1	1	1	1	1	1	2	1	1	-	1	1	1	2	-	2	2	1	1	1	2	3	3	3	2
Diaphanosoma brachyurum	-	-	-	-	-	-	-	-	-	-	-	-	1	1	1	-	-	-	1	1	1	1	1	1	1	-	1	1	1	1	1	2	-	1	1	3	1	1	1	3	3	3	1
Diaptomus	-	-	-	-	-	-	-	-	-	-	-	1	1	1	1	1	1	1	1	1	-	-	1	-	-	-	-	-	1	1	2	2	2	2	1	1	-	1	1	2	2	2	1
Daphnia	-	-	-	-	-	-	-	-	-	-	-	-	-	-	-	-	-	-	-	-	-	-	-	-	-	-	-	-	-	-	-	-	-	-	-	-	-	-	-	-	-	-	-
Harpacticida	-	-	-	-	-	-	-	-	-	-	-	-	-	-	-	-	-	-	-	-	-	-	-	-	-	-	-	-	-	-	-	-	-	-	-	-	-	-	-	-	-	-	-
Iliocryptus	-	-	1	-	-	1	1	-	1	-	-	1	1	1	-	1	1	1	-	1	1	-	-	-	-	-	-	-	-	-	-	-	1	-	-	-	-	-	-	-	-	-	-
Moina	-	-	-	-	-	-	-	-	-	-	-	1	1	1	1	1	1	-	-	-	-	-	-	-	-	-	-	-	-	-	-	-	-	-	-	-	-	-	-	-	-	-	-
Ostracoda	-	-	-	-	-	-	1	-	1	-	-	-	-	-	-	-	-	-	-	-	-	-	-	-	-	-	-	-	-	-	-	-	-	-	-	-	-	-	-	-	-	-	-
R o t a t o r i a																																											
Amureopsis fissa	1	-	1	-	-	-	1	-	-	-	-	1	1	1	-	-	-	-	-	-	-	-	-	-	-	-	-	-	-	-	-	-	-	-	-	-	-	-	-	-	-	-	-
Ascomorpha saltans	-	-	-	-	-	-	-	-	-	-	-	1	-	1	-	-	-	-	-	-	-	-	-	-	-	-	-	-	2	1	-	-	-	-	-	-	-	1	3	2	1	3	1
Asplanchna	-	-	1	1	-	-	1	-	-	-	-	1	2	-	-	-	1	-	-	-	1	-	-	-	1	1	1	-	2	1	1	-	-	-	-	-	-	-	1	1	1	1	1
Brachionus calyciflorus	-	-	-	-	-	-	-	-	-	-	-	1	-	-	-	-	-	-	-	-	-	-	-	-	-	-	-	-	-	-	-	-	-	-	-	-	-	-	1	1	1	1	1
Brachionus havaniensis	1	-	1	-	1	1	1	1	-	-	-	1	-	-	-	1	1	1	-	1	1	1	1	1	1	1	1	1	1	1	1	-	-	-	-	-	-	-	-	-	-	-	-
Brachionus quadratus	-	-	-	-	-	-	-	-	-	-	-	1	-	-	-	-	-	-	-	-	-	-	-	-	2	-	-	-	-	-	-	-	-	-	-	-	-	-	-	-	-	-	-
Cathypna luna	1	1	1	-	1	1	1	1	1	1	1	1	1	1	1	1	1	1	-	1	1	1	-	1	-	-	-	-	1	1	1	-	1	1	1	-	-	-	-	-	-	-	-
Cephalodella	1	1	1	1	1	-	1	1	1	1	-	1	-	-	-	-	-	-	-	-	-	-	1	1	-	-	-	-	-	-	-	-	2	2	2	2	-	-	-	-	-	-	-
Collotheca mutabilis	-	-	-	-	-	-	-	-	-	-	-	1	2	-	-	-	1	1	1	-	-	-	-	-	-	-	-	-	-	-	-	-	-	-	-	-	-	-	-	-	-	-	-
Conochiloides coenobasis	-	-	-	-	-	-	-	-	-	-	-	1	1	2	-	-	1	-	-	-	1	-	-	-	2	1	1	-	1	2	1	1	1	1	2	1	-	1	1	1	1	1	1
Dipleuchlanis	1	-	-	-	-	-	-	-	1	-	-	-	-	-	-	-	-	1	-	1	-	-	-	-	2	1	1	-	1	2	1	1	2	1	1	2	5	3	1	1	1	1	1
Euchlanis	1	-	-	-	1	-	1	-	-	-	-	1	1	1	-	1	1	1	1	1	1	1	1	1	-	-	-	-	-	-	-	-	-	-	-	-	-	-	-	-	-	-	-
Eudactylota eudactylota	-	-	-	-	-	-	1	-	-	-	-	-	-	-	-	-	1	1	1	1	1	1	1	1	1	1	1	1	2	1	1	-	-	-	-	-	-	-	-	-	-	-	-
Filinia longiseta	1	1	1	1	-	-	1	-	-	1	-	1	1	-	1	1	1	1	1	-	-	-	1	-	1	1	1	-	1	1	1	-	-	-	-	-	-	-	-	-	-	-	-
Hexarthra insulina	1	1	1	1	-	-	-	-	-	-	-	1	-	1	-	1	1	1	1	1	1	1	1	1	-	-	-	-	1	1	1	1	-	-	-	-	-	-	-	-	-	-	-
Keratella americana	-	-	-	-	-	-	1	1	1	1	1	1	1	1	1	1	1	1	1	1	1	1	1	1	1	1	1	1	1	1	1	1	1	1	1	1	1	1	1	1	1	1	1
Lecane ludwigi	-	-	-	-	-	-	1	-	-	-	-	-	1	-	1	1	1	1	1	1	1	-	1	-	-	-	-	-	-	-	-	-	-	-	-	-	-	-	-	-	-	-	-
Lecane quadridentata	2	1	1	1	1	-	1	1	1	1	-	1	1	1	1	1	1	1	1	1	1	-	1	1	1	1	1	1	1	1	1	1	-	-	-	-	1	1	-	-	-	-	-
Lecane indeterm.	1	-	-	-	-	-	1	-	-	-	1	1	-	-	-	1	1	1	-	1	-	-	-	-	-	-	-	-	-	-	-	-	-	-	-	-	-	-	-	-	-	-	-
Lepadella patella	-	-	-	-	-	-	-	-	-	-	-	-	-	-	-	-	-	1	-	1	1	-	1	-	1	1	-	1	1	1	-	1	1	-	-	-	-	-	-	-	-	-	1
Lepadella indeterm.	-	-	1	-	1	-	-	1	1	-	-	-	-	-	-	1	1	1	-	1	1	1	-	-	1	1	1	-	-	-	-	-	-	-	-	-	-	-	-	-	-	-	-
Monommata longiseta	-	-	-	-	-	-	-	-	1	-	-	-	-	-	-	1	1	1	-	1	1	1	-	-	-	-	-	-	1	1	-	-	-	-	-	-	-	-	-	-	-	-	-
Monostyla bulla	-	-	-	-	-	-	-	-	-	-	-	-	-	-	-	1	1	-	1	1	-	-	-	-	1	1	1	-	1	1	1	1	1	1	1	-	-	-	-	-	-	-	-
Platyias patulus	-	-	-	-	-	-	1	1	1	1	1	1	1	1	1	1	1	1	1	1	1	1	1	1	2	1	1	1	1	1	1	1	1	1	1	1	1	-	-	-	-	-	-
Platyias quadricornis	1	1	1	1	1	1	1	1	1	1	1	2	1	1	1	1	1	1	1	1	1	1	1	1	2	2	1	1	1	1	1	1	1	1	1	1	1	1	1	-	-	-	1
Polyarthra	1	1	1	1	-	1	2	1	1	-	-	2	1	1	1	1	1	-	-	1	-	-	-	-	1	1	1	-	1	1	1	-	-	-	-	-	-	-	-	-	-	-	-
Rotaria	-	-	-	-	-	-	-	-	-	-	-	-	-	-	-	1	1	-	-	1	-	-	-	-	2	2	1	1	-	-	-	1	-	-	-	-	-	1	1	1	-	-	-
Salpina macracantha	-	1	1	-	-	-	1	-	-	-	-	1	1	1	1	-	1	1	-	1	2	1	1	2	1	1	1	1	1	1	1	1	2	2	1	2	-	-	1	-	-	-	-
Sinantherina spinosa	-	-	-	-	-	-	-	-	-	-	-	-	-	-	-	2	-	-	-	3	1	2	1	-	-	-	-	-	-	-	-	-	-	-	-	-	-	-	-	-	-	-	-
Synchaeta	2	-	1	1	-	1	1	-	1	1	1	1	1	1	-	1	1	1	-	1	1	-	-	-	1	1	-	-	-	1	-	1	-	2	1	2	-	-	1	-	-	-	-
Testudinella patina	2	-	-	-	-	-	-	-	-	-	-	1	-	-	-	-	-	-	-	-	-	-	-	-	-	-	-	-	1	-	-	1	-	-	-	-	-	-	-	1	1	-	-

166

Table XIII.   (1st. continued)

Trichocerca similis	-	-	-	-	-	-	-	-	-	-	-	1	1	1	-	-	-	-	-	-	-	-	-	-	-	-	-	-	-	-	-	-	-	-	-	-	-	-	-	-	-	-	-
indeterm.	-	1	-	1	-	-	1	-	-	-	-	-	-	-	-	3	2	-	-	-	-	-	-	-	-	-	-	-	-	-	-	-	-	-	-	-	-	-	-	-	-	-	-
Protozoa																																											
Amoeba	-	-	1	1	-	-	-	-	-	-	-	-	-	-	-	-	-	-	-	-	-	-	-	-	-	-	-	-	-	-	-	-	-	-	-	-	-	-	-	-	-	-	-
Arcella	1	1	1	1	1	1	1	1	1	1	1	1	2	1	-	1	2	1	-	1	-	-	1	2	-	-	-	-	-	-	-	-	-	-	1	1	1	-	-			
Coleps hirtus	-	-	-	-	-	-	-	-	-	-	-	-	-	-	-	-	-	-	-	2	-	-	-	-	-	-	-	-	-	-	-	-	-	-	-	-	-	-	-			
Difflugia	1	1	1	1	-	1	1	1	1	1	1	-	-	-	1	1	1	-	1	-	-	1	-	-	-	-	-	-	-	-	-	-	1	2	1	2	-	-	-			
Euglypha	-	-	-	-	1	1	2	1	-	1	-	-	-	-	-	-	-	-	-	-	-	-	-	-	-	-	-	-	-	-	-	-	-	-	-	-	-	-				
Heliozoa	3	-	-	-	-	1	-	1	-	2	-	-	-	-	-	-	-	-	-	-	-	-	-	-	-	-	-	-	-	-	-	-	-	-	-	-	-				
Rhipidodendron huxleyi	-	-	2	1	-	-	-	-	-	-	-	-	-	-	-	-	-	-	-	-	-	-	-	-	-	-	-	-	-	-	-	-	-	-	-	-	-				
Tintinnidium	-	-	1	-	-	-	-	-	1	-	-	-	-	-	-	-	-	-	-	-	-	-	-	-	-	-	-	-	-	-	-	-	-	-	-	-					
Tokophrya?	-	-	1	1	-	-	-	1	-	-	1	-	-	-	-	-	-	-	-	-	-	-	-	-	-	-	-	-	-	-	-	-	-	-	-					
Flagellata																																											
Dinobryon soc.var.americana	-	-	-	1	-	-	-	-	-	-	-	-	-	-	-	-	-	-	-	-	-	-	-	-	-	-	-	-	-	-	-	-	-	-	-						
Eudorina elegans	1	1	-	1	-	-	-	1	-	1	2	1	1	-	1	2	3	1	2	1	1	1	1	-	1	-	1	2	2	1	1	-	-	1	-	-				
Euglena	-	-	1	1	1	1	1	-	1	1	1	1	-	-	-	-	-	-	1	1	-	1	-	1	-	1	-	1	-	-	-	-	-	-	-					
Gonium pectorale	1	-	-	-	-	-	-	-	-	-	-	-	-	-	-	-	-	-	-	-	-	-	-	-	-	-	-	-	-	-	-	-	-	-	-					
Lepocinclis	-	-	-	-	-	-	-	-	-	-	-	-	-	-	-	-	-	2	1	-	2	-	-	-	-	-	-	-	-	-	-	-	-	-						
Mallomonas	-	-	-	-	-	-	-	-	-	-	-	-	-	-	-	-	-	-	-	1	-	-	-	-	-	-	-	-	-	-	-	1	1	-						
Pandorina morum	1	1	1	-	1	1	1	-	1	-	-	-	-	-	1	1	1	1	1	-	-	-	-	-	-	-	-	-	-	-	-	-	-	-						
Peridinium	-	-	-	-	-	-	-	-	-	-	1	-	-	-	-	-	-	-	2	-	1	-	1	-	-	-	-	-	-	-	-	-	-	-						
Phacus longicauda	-	-	-	-	-	-	-	-	-	1	-	-	-	-	-	-	1	-	1	2	-	-	-	-	-	-	-	-	-	-	-	-	-							
Phacus indeterm.	-	-	-	-	-	1	-	-	-	-	-	-	-	1	1	1	1	2	-	-	1	-	-	-	-	-	-	-	-	-	-	-							
Pyrobotrys	-	-	-	-	-	-	-	-	-	-	-	2	1	1	1	1	2	-	-	1	-	-	-	-	-	-	-	-	-	-	-								
Strombomonas ensifera	-	-	-	-	-	-	-	-	-	-	-	-	-	-	-	-	1	2	-	-	-	-	-	-	-	-	-	-	-	-								
Synura uvella	-	-	-	-	-	-	-	-	-	-	-	-	-	1	-	-	-	-	-	-	-	-	-	-	-	-	-	-	-									
Trachelomonas caudata	-	1	1	1	-	1	2	1	1	-	1	1	-	-	-	1	-	-	1	-	-	-	-	-	-	-	-	-	-	-									
Trachelomonas indeterm.	-	-	-	-	-	-	-	-	-	-	-	1	1	2	2	1	2	3	3	1	1	1	-	2	1	1	-	1	1	1	2	2	1
Uroglena americana	-	-	-	-	-	-	-	1	1	-	-	-	-	-	-	-	-	-	1	-	-	-	-	-	-	-	-										
indeterm.	-	1	-	-	-	2	2	1	1	1	-	-	1	2	-	1	2	2	2	2	2	-	3	2	3	2	1	1	1	2	2	2	2	2	3	2	1	1	1	1	-
Chlorophycea																																											
Ankistrodesmus falcatus	-	-	-	-	-	-	-	-	-	-	-	-	-	-	-	-	-	-	-	-	-	-	-	-	-	-	-	1	-	-	-	-	1	-						
Coelastrum cambricum	-	1	-	-	-	-	-	-	-	-	-	-	-	-	-	-	-	-	-	-	-	-	-	-	-	-	-	-	-	-	-	-								
Crucigenia rectangularis	-	1	-	-	-	-	-	-	-	-	-	-	-	-	-	-	-	-	-	-	-	-	1	1	-	1	-	1	-	-	-	-								
Dictyosphaerium pulchellum	-	-	-	-	-	-	-	-	2	-	2	1	-	-	-	1	1	3	2	1	2	-	1	1	1	2	1	2	1	-	1	1	1	1	1	3	3	3	1
Dimorphococcus lunatus	-	-	-	-	-	-	-	1	-	-	1	-	-	1	-	-	1	-	1	-	1	-	1	2	-	-	1	-	-	-	-								
Franceia?	-	-	-	-	-	-	1	-	-	-	-	-	-	-	-	-	-	-	-	-	-	-	-	-	-	-	-	-											
Kirchneriella obesa	-	-	-	-	-	-	-	1	-	-	-	-	-	-	2	-	1	1	-	1	-	1	1	-	-	-	1	-	1	3	-	-								
Microspora?	1	-	-	-	-	-	-	-	-	1	-	2	-	-	-	-	-	-	-	-	-	-	-	-	-	-	-	-	-	-	2	-								
Mougeotia	-	-	-	-	-	-	-	-	-	-	-	-	-	-	-	-	-	-	-	-	-	-	-	-	-	-	1	-	-	-	1	-								
Pediastrum tetras	-	1	-	-	-	-	-	-	-	-	-	-	-	-	-	-	-	-	-	-	-	-	-	-	-	-	-	-	-	-									
Scenedesmus quadricauda	1	-	1	-	-	1	-	1	-	r	1	-	-	-	-	-	-	-	-	-	-	-	-	-	-	1	-	-	-	-									
Spirogyra	-	-	1	1	-	-	1	1	1	-	1	1	1	-	1	1	1	1	-	-	-	-	-	-	-	-	-	-	-	-									
Tetraedron	1	1	-	-	-	-	-	-	-	-	-	-	-	-	-	-	-	-	-	-	-	-	-	-	-	-	-	-	-	-	-									
indeterm. green colonies	-	-	-	-	-	-	-	-	-	-	-	-	-	-	-	-	2	3	-	-	-	1	-	1	-	-	3	3	3	3	3	3
indeterm. green filaments	-	-	-	-	-	-	1	-	-	-	-	-	1	1	1	1	-	-	-	-	-	-	-	1	-	-	-	-	-	-	-								

Table XIII. (2nd. continued)

D e s m i d i a c e a e
Closterium kuetzingii
Closterium indeterm.
Cosmarium
Cosmarium indeterm.
Cosmocladium
Docidium
Euastrum
Gonatozygon
Hyalotheca mucosa
Micrasterias arcuatus
Micrasterias indeterm.
Penium margaritaceum
Sphaerozosma granulatum
Staurastrum inaequale
Staurastrum mamillatum
Staurastrum trifidum
Staurastrum indeterm.
Xanthidium antilopeum
D i a t o m e a e
Cyclotella
Diatoma
Eunotia asterionelloides
Melosira granulata
Navicula
pennatae
Pinnularia
Rhizosolenia
Surirella
Synedra
C y a n o p h y c e a e
Lyngbya
Oscillatoria splendida
Oscillatoria indeterm.
indeterm. filaments

Actinospora(Fungi imperfecti)
cercariae (unifurcated)
cercariae (bifurcated)
Chaetogaster
Ephemerida
Gastrotricha
Hydracarinae
iron bacteria indeterm.
Leptothrix
mosquito larvae
Nematoda
Odonata
Plumatella zooecia
Pristina longiseta
sponges spicula
veliger

Table XIV. Plankton from Beerdotti to Afobaka (29/30. IV. 1964)

	Beerdotti	Geltoesi	Gansee	Lombe	Saidagoe	Kabel-S	Kabel	Kadjoe	Aloesoebanja-S	Aloesoebanja-N	Koffiekamp	Afobaka
C r u s t a c e a												
Bosmina	1	-	-	-	1	-	2	1	1	1	-	-
Bosminopsis	-	-	-	-	1	1	-	-	-	-	-	-
Ceriodaphnia	-	-	-	-	-	-	1	1	3	2	-	-
Chydorus	-	-	-	-	-	-	-	1	1	-	1	-
Cyclops	-	-	-	-	1	1	-	1	2	2	1	1
Diaphanosoma	-	-	-	-	-	-	1	1	2	1	1	-
Diaptomus	-	-	-	-	-	-.	-	-	1	-	-	-
Moina	-	-	-	-	1	-	-	-	-	1	-	1
R o t a t o r i a												
Anureopsis	-	1	-	-	-	-	-	-	-	-	-	-
Asplanchna	-	-	-	-	-	-	-	-	-	-	-	1
Brachionus quadridentata	-	-	-	-	-	-	-	-	1	1	1	-
Collotheca	-	1	1	-	1	1	1	-	-	-	-	-
Conochiloides coenobasis	-	1	1	1	-	-	-	1	1	2	1	2
Filinia longiseta	-	-	-	-	-	-	-	1	1	1	2	2
Keratella americana	-	1	1	1	1	-	-	-	-	-	-	-
Pedalion	-	1	1	2	2	1	1	1	1	2	-	-
Polyarthra	1	2	2	2	2	2	1	1	1	1	1	1
Sinantherina	-	-	-	-	-	-	-	-	1	-	-	-
Testudinella brycei	-	-	1	-	-	-	-	-	-	-	-	-
P r o t o z o a												
Arcella	-	1	-	1	-	1	-	-	-	-	-	-
Bursaria	-	-	-	-	-	-	-	-	-	-	3	-
Difflugia	1	-	-	-	-	-	-	-	-	-	-	-
Heliozoa	2	2	1	1	1	1	-	-	-	-	-	-
Microstomum	-	-	-	-	-	-	-	-	-	-	3	-
F l a g e l l a t a												
Dinobryon	-	-	-	1	-	-	-	-	-	-	-	-
Euglena	-	-	-	-	-	1	1	2	2	2	-	-
Eudorina	1	1	1	1	1	2	2	2	1	1	-	-
small flagellates	-	-	-	-	-	2	1	2	2	2	2	-
Gonium pectorale	-	-	1	1	-	-	-	-	-	-	-	-
Peridinium	-	1	1	1	1	1	1	1	-	-	-	-
Pyrobotrys	-	-	-	-	-	-	-	-	-	-	1	1
Strombomonas	-	-	-	-	-	-	1	3	3	3	1	1
Synura	-	-	1	1	1	1	-	-	1	1	-	-
Trachelomonas caud.	-	-	-	-	1	1	-	-	-	-	-	-
C h l o r o p h y c e a												
Dictyosphaerium	-	2	2	2	1	1	1	1	2	1	1	-
Dimorphococcus	-	-	-	-	-	-	-	-	-	1	1	1
Franceia ?	1	1	1	1	1	2	1	-	-	-	-	-
Kirchneriella	-	-	1	-	-	1	-	1	-	1	1	-
Mougeotia	1	1	1	1	1	-	-	-	-	-	-	-
Scenedesmus bijuga	-	-	-	1	-	-	1	1	-	-	1	-
Spirogyra	-	1	-	-	-	-	-	-	-	-	-	-
D e s m i d i a c e a e												
Closterium setaceum	1	-	-	-	-	-	-	-	1	1	1	1
Cosmarium	1	1	1	1	1	-	-	-	-	-	-	-
Cosmocladium	-	-	1	1	-	-	1	1	3	2	2	1
Holacanthum	-	-	-	-	-	-	-	-	-	1	-	-
Hyalotheca	-	-	-	-	-	-	-	1	-	1	1	-
Schizacanthum ?	1	2	1	1	2	1	1	1	1	-	-	-
Sphaerozosma	1	1	1	1	2	2	-	-	-	-	-	-
Staurastrum mamillatum	1	1	1	1	1	2	1	2	2	2	2	1
Staurastrum inaequale	-	1	1	1	2	1	1	1	1	1	1	1
Staurastrum minnesotense	1	-	-	-	1	-	-	1	-	1	1	
Staurastrum	-	1	1	1	1	1	1	1	2	2	1	1
D i a t o m e a e												
Eunotia asterionelloides	4	4	4	4	4	4	3	2	1	-	-	-
Cyclotella	-	-	-	-	-	-	-	1	1	1	1	-
Melosira	-	-	-	-	-	1	-	1	1	1	-	-
Rhizosolenia	1	-	1	-	-	1	-	-	-	-	-	-
Surirella	1	-	-	-	-	-	-	-	-	-	-	-
sponges (spicula)	1	1	1	1	-	-	-	-	-	-	-	-
veliger	1	1	1	-	-	-	-	-	-	-	-	-

170

Table XV. Sedimentation plankton in 1 liter water from Station Afobaka (16. IV. 1968)

1 = present	3 = many specimens	5 = very abundant	
2 = few specimens	4 = abundant	( )= no living specimens	

depth in m:	0	5	10	15	20	25	30	35
C r u s t a c e a								
Bosmina	-	1	-	-	-	-	-	1
Cyclops	1	1	-	1	-	1	-	-
Diaphanosoma	-	1	-	-	-	-	-	-
Ostracoda	1	-	-	-	-	-	-	1
R o t a t o r i a								
Anureopsis fissa	-	1	-	-	-	-	-	-
Filinia longiseta	-	-	-	-	1	-	-	-
Keratella americana	1	-	-	-	-	-	-	-
Trochosphaera	-	-	1	-	-	-	-	-
P r o t o z o a								
Arcella	-	-	2	-	-	-	-	-
Centropyxis	-	-	-	-	-	1	-	-
F l a g e l l a t a								
Dinobryon	1	1	-	-	-	-	-	-
Eudorina	2	1	1	1	1	-	1	-
Peridinium	2	1	-	1	1	1	1	1
Trachelomonas	2	1	1	1	-	-	1	1
Volvox	1	1	-	1	-	-	1	-
unicellular	3	2	2	2	-	-	-	-
indeterm. colonies	3	-	-	-	-	-	-	-
bacteria ?	-	-	1	2	1	2	3	3
C h l o r o p h y c e a								
Ankistrodesmus	3	3	-	1	2	2	-	-
Oegodonium	-	-	-	-	-	3	-	-
Scenedesmus	1	-	-	-	-	-	-	-
fungi	1	1	1	1	1	1	3	3
D e s m i d i a c e a e								
Closterium kuetzingii	2	1	1	-	1	1	-	-
Cosmarium	2	2	-	-	-	-	-	-
Desmidium	-	-	-	-	-	-	-	1
S taurastrum leptacanthum	3	3	2	2	2	2	-	(1)
Staurastrum leptocladum	1	1	1	1	-	1	-	-
Staurastrum sp.	3	3	2	2	1	-	1	-
Staurastrum sp.	2	2	1	-	1	1	1	(1)
D i a t o m e a e								
pennatae	-	-	1	-	-	-	-	-

Table XVI. Sedimentation plankton in 1 liter water from Station Sara (17. IV. 1968)

| 1 = present | 3 = many specimens | 5 = very abundant |
| 2 = few specimens | 4 = abundant | ( ) = no living specimens |

depth in m:	0	5	10	15	20	25	30
C r u s t a c e a							
Diaphanosoma	-	-	1	-	-	-	-
Diaptomus	1	-	-	-	-	-	-
R o t a t o r i a							
Brachionus falcatus	1	-	-	-	-	-	-
Collotheca	1	-	-	-	-	-	-
Hexarthra	1	-	-	-	-	-	-
Polyarthra	1	-	-	-	-	-	-
Sinantherina	1	-	-	-	-	-	-
Synchaeta	-	1	-	-	-	-	-
P r o t o z o a							
Euglypha	-	-	-	-	-	-	(1)
F l a g e l l a t a							
Dinobryon	-	1	-	-	-	-	-
Eudorina	3	1	1	-	(1)	-	-
Peridinium	2	1	-	-	-	-	-
Trachelomonas	1	1	1	-	-	-	-
Volvox	3	1	1	-	-	-	1
indeterm. colonies	3	2	1	-	-	1	1
unicellular	2	1	1	-	1	-	1
C h l o r o p h y c e a							
Ankistrodesmus	1	1	-	-	-	-	-
D e s m i d i a c e a e							
Closterium kuetzingii	1	1	-	-	-	-	-
Onychonema	-	-	-	-	(1)	-	-
Staurastrum dejectum	3	3	2	-	(1)	(1)	-
Staurastrum leptacanthum	3	3	2	-	(1)	(1)	(1)
Staurastrum leptocladum	1	1	1	-	(2)	(2)	(2)
Staurastrum subgrande ?	2	1	1	-	(1)	(2)	-
D i a t o m e a e							
Melosira	-	-	-	-	(1)	(1)	-
Melosira-fragments	-	-	-	-	(2)	(3)	(2)
pennatae	1	1	1	-	1	(1)	(2)

Table XVII. Net-plankton from Brokopondo Lake (April 1968)

1 = present   3 = many specimens   5 = very abundant
2 = few specimens   4 = abundant

	Afobaka 30.IV. green	Kabel 30.IV. green	Bedotti 30.IV. green	near Grankreek 25.IV. green	Panpantiri 25.IV. greenish	Adewai 25.IV. pale green	Mammadam 24.IV. pale green	Pokigron 23.IV. brown	Sara 30.IV. brown	Locus 30.IV. brown	Sikakoempoe 12.IV. greenish brown	Station 8 27.IV. green	Station 8-Red. 27.IV. green	Redidotti 27.IV. green	railroad Kabel II.IV. brownish green
Crustacea															
Alona sp. 1	-	-	-	-	1	2	-	-	-	-	-	-	-	-	-
Alona sp. 2	-	-	-	-	1	1	-	-	-	-	-	-	-	-	-
Alona sp. 3	-	-	-	-	1	-	-	-	-	-	-	-	-	-	-
Argulus	-	-	1	-	-	1	-	-	-	-	-	-	-	-	-
Bosmina	1	-	1	-	-	-	-	-	1	1	-	1	-	2	1
Bosminopsis	-	-	-	-	1	1	-	1	-	-	-	-	-	-	-
Ceriodaphnia cornuta	2	1	2	5	3	1	-	-	3	3	1	1	1	3	2
Cyclops	2	1	2	5	3	-	-	-	2	2	1	1	1	2	2
Daphnia longispina	1	-	-	-	-	-	-	-	1	1	-	-	1	1	1
Diaphanosoma brachyurum	2	2	2	3	1	-	-	-	2	2	1	3	1	-	2
Diaptomus	2	1	1	1	-	1	-	-	3	3	1	4	2	3	3
Harpaticida	-	-	-	-	-	-	1	1	-	-	-	-	-	-	-
Ostracoda	2	1	3	-	-	1	-	-	3	3	2	4	1	4	3
Polyphemus pediculis	-	-	-	-	-	1	-	-	-	-	-	-	-	-	-
Rotatoria															
Asplanchna	-	1	1	-	1	-	-	-	1	-	-	-	-	-	-
Brachionus falcatus	1	1	1	-	1	-	-	-	1	2	-	1	-	-	-
Collotheca	2	2	1	2	1	-	-	-	3	3	-	1	1	-	-
Conochiloides coenobasis	-	-	1	3	2	-	-	1	-	1	1	1	-	1	2
Euchlanis	-	-	-	-	-	1	-	-	-	-	-	-	-	-	-
Hexarthra	-	-	1	1	3	-	-	-	1	1	1	1	-	-	1
Keratella americana	-	-	1	1	1	-	-	-	-	1	-	1	-	-	-
Lecane	-	-	-	-	2	1	-	-	-	1	-	-	-	-	-
Macrochaetus	-	-	-	-	-	1	-	-	-	-	-	-	-	-	-
Platyias patulus	-	-	-	-	-	1	-	-	-	-	-	-	-	-	-
Polyarthra	-	-	1	2	1	-	-	-	-	-	1	1	1	1	-
Rotaria	-	-	-	-	-	-	1	-	-	-	-	-	-	-	-
Sinantherina spinosa	-	-	1	-	-	-	-	-	-	-	-	2	-	1	1
Testudinella patina	-	-	-	-	-	1	-	-	-	-	-	-	-	-	-
Trichocerca	1	-	1	1	-	-	-	-	1	-	-	-	-	-	1
Trochosphaera	1	1	-	-	-	-	-	-	1	2	-	-	-	-	-
Protozoa															
Arcella	-	-	-	-	1	-	1	1	-	-	-	-	-	-	-
Centropyxis aculeata	-	-	1	-	-	-	-	-	-	-	-	-	-	-	-
Difflugia	-	-	-	-	-	-	-	1	-	-	-	-	-	-	-
Lesquereusia	-	-	-	-	-	-	-	-	-	1	-	-	-	-	-
Flagellata															
Dinobryon	-	-	1	3	-	-	-	-	-	-	-	1	1	1	-
Eudorina elegans	2	2	3	4	4	1	1	1	3	3	3	2	3	2	3
Peridinium	-	-	1	-	-	-	-	-	-	-	-	-	-	-	-
Rhipidodendron huxleyi	-	-	-	-	-	1	1	2	-	-	-	-	-	-	-
Trachelomonas	-	-	-	-	-	-	-	-	-	-	-	-	-	1	-
Volvox	2	1	2	3	3	1	-	-	2	2	2	3	3	2	5
unicellular	-	-	-	3	3	2	-	-	-	-	-	-	-	-	-

Table XVII. (continued) Net-plankton from Brokopondo Lake (April 1968)

	Afobaka 30.IV. green	Kabel 30.IV. green	Bedotti 30.IV. green	near Grankreek 25.IV. green	Panpantiri 25.IV. greenish	Adewai 25.IV. pale green	Mammadam 24.IV. pale green	Pokigron 23.IV. brown	Sara 30.IV. brown	Locus 30.IV. brown	Sikakoempoe 12.IV. greenish brown	Station 8 27.IV. green	Station 8-Red. 27.IV. green	Redidotti 27.IV. green	railroad Kabel 11.IV. brownish green
Chlorophycea															
Ankistrodesmus falcatus	-	-	-	-	-	-	-	-	-	-	1	-	-	1	-
Botryococcus brauni	1	1	2	3	-	-	-	-	1	1	2	3	3	2	2
Cladophora ?	-	-	-	-	-	-	-	-	-	-	1	-	-	-	-
Dictyosphaerium	1	-	1	1	-	-	-	-	-	-	-	-	1	2	-
Dimorphococcus lunatus	-	-	-	-	-	-	-	-	-	-	-	-	-	-	1
Mougeotia	-	-	-	-	-	-	-	-	-	-	-	-	-	1	1
Selenastrum gracile	-	-	-	-	-	-	-	-	-	-	-	-	-	-	1
Spirogyra	-	-	-	-	-	-	-	-	-	-	1	-	-	-	-
Desmidiaceae															
Closterium kuetzingii	1	1	2	2	2	1	-	-	1	1	1	2	1	2	1
Closterium	-	-	-	-	-	-	-	-	-	-	2	-	-	-	-
Cosmarium	2	2	3	4	4	-	-	-	2	2	2	3	2	1	1
Cosmocladium	2	3	2	2	-	-	-	-	2	2	-	-	2	-	-
Desmidium	-	-	-	-	-	-	-	-	-	-	-	-	-	-	2
Hyalotheca	-	-	-	-	-	1	-	-	-	-	1	1	-	-	2
Micrasterias brasiliensis	-	-	1	-	-	-	-	-	-	1	-	-	-	1	3
Micrasterias sp. 1	-	-	-	-	-	-	-	-	-	-	-	-	-	1	-
Micrasterias sp. 2	-	-	-	-	-	-	-	-	-	-	-	-	-	1	-
Onychonema	-	-	1	-	-	-	-	-	-	1	-	-	-	-	3
Staurastrum leptacanthum	5	5	2	3	2	1	1	1	5	5	3	2	2	1	1
Staurastrum leptocladum	-	-	1	3	1	-	-	-	-	-	1	1	-	1	2
Staurastrum mahabulaswarensis	-	-	-	-	-	-	-	-	-	-	-	-	-	-	1
Staurastrum subgrande ?	4	4	1	-	-	-	-	-	3	4	4	5	5	-	1
Staurastrum	-	-	-	2	-	-	-	-	-	-	-	-	-	1	-
Xanthidium antilopeum	-	-	-	1	1	-	-	-	-	-	-	-	-	1	2
indeterm.	-	-	3	-	-	-	-	-	-	-	-	-	2	-	-
Diatomeae															
Surirella	-	-	-	-	-	-	1	-	-	-	-	-	-	-	-
Pinnularia viridis	-	-	-	-	-	-	-	1	-	-	-	-	-	-	-
Cyanophyceae															
Oscillatoria	-	-	-	-	-	-	-	-	-	-	-	1	-	-	-
Gloeocystis ?	-	-	-	-	-	-	-	-	-	-	-	-	-	2	-
Bryozoa	-	-	1	-	1	-	-	-	-	-	-	-	-	-	2
Chaoboris	-	-	-	-	-	-	-	-	-	-	1	-	-	-	2
dragonfly egg	-	-	-	-	-	-	1	-	-	-	-	1	-	-	-
Gasterotricha	-	-	-	-	-	-	-	-	-	-	-	1	-	-	-
ironbacteria	-	-	-	-	-	1	1	-	-	-	-	-	-	-	-
Nematoda	-	-	-	-	-	-	1	-	-	-	-	-	-	-	-
Podostemonaceae fragments	-	-	-	-	-	1	1	-	-	-	-	-	-	-	-
sponges spiculae	-	-	-	-	-	2	2	-	-	-	-	-	-	-	-

PLATE I

Pl. I. Aerial view of the Afobaka area, one and a half year after the closing of the dam, looking southwest [June 1965].

The photographs of Pls. I, VII, VIIIb and IXb were made available by CENTRAAL BUREAU LUCHTKARTERING SURINAME; Pls. IIa, VIIIa and IXa by P. WAGENAAR HUMMELINCK. All microphotographs on Pls. X–XVIII are taken by Mrs. H. MEYER.

PLATE II

Pl. IIa. Aloesoebanja rapids, south of Afobaka, with flowering Podostemaceae (*Mourera fluviatilis*) [Sep. 1955].
Pl. IIb. Aloesoebanja rapids, south of Afobaka, 8 months after the closing of the dam [Sep. 1964].

PLATE III

Pl. IIIa. Suriname River, as seen from the dam downwards, a few hours after the closing of the dam [1.II.1964].

Pl. IIIb. Suriname Grandam rapids in upper course of Suriname River near Ligolio, Gran Rio, with flowering *Mourera* [Aug. 1964].

PLATE IV

Pl. IVa. Saramacca River near Mamadam [Apr. 1964].
Pl. IVb. Grankreek, a tributary of the Suriname River [Aug. 1964].

PLATE V

Pl. Va. Waterfern (*Ceratopteris pteridioides*) and duckweed (*Lemna valdiviana*) in the Sara Kreek, 6 weeks after the closing of the dam [Apr. 1964].
Pl. Vb. Dense film of iron bacteria in the Sara Kreek, 6 weeks after the closing of the dam [Apr. 1964].

PLATE VI

Pl. VIa. Mats of filamentous algae (with gas bubbles) and duckweed in the Broko-
pondo Lake near Koffiekamp [July 1964].
Pl. VIb. Waterhyacinth (*Eichhornia crassipes*) in the Brokopondo Lake near Kabel
[July 1964].

PLATE VII

Pl. VII. Aerial view of Brokopondo Lake, showing growths of *Eichhornia* around dead trees [Nov. 1966].

PLATE VIII

Pl. VIIIa. Border of Brokopondo Lake at Locus Kreek [Oct. 1968].
Pl. VIIIb. Aerial view of Brokopondo Lake, with aeroplane (upper-left) spraying
chemicals during waterhyacinth campaign [Nov. 1966].

PLATE IX

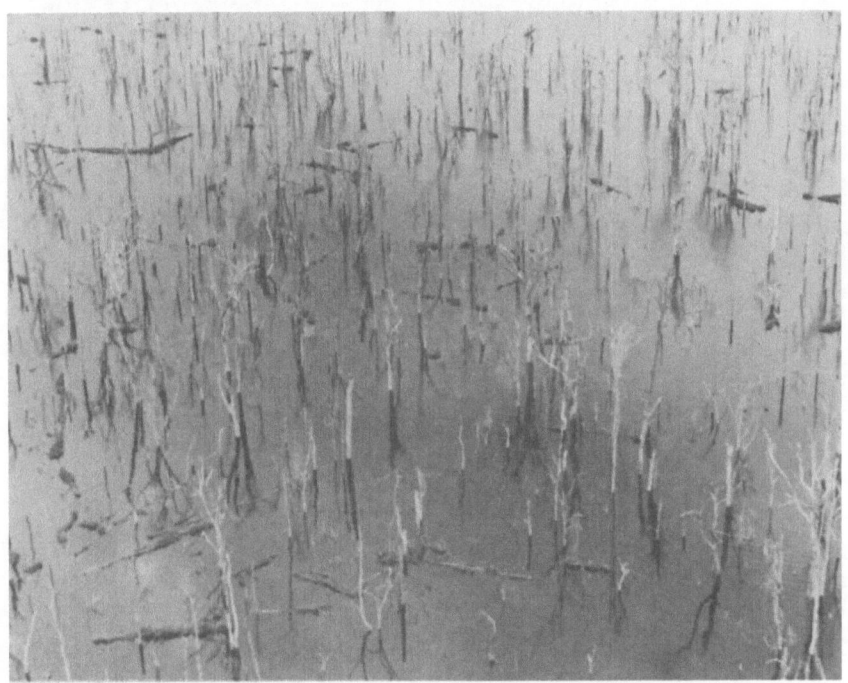

Pl. IXa. Northern part of Brokopondo Lake [Oct. 1968].
Pl. IXb. Aerial view of Brokopondo Lake with trunks of dead trees [July 1969].

PLATE X

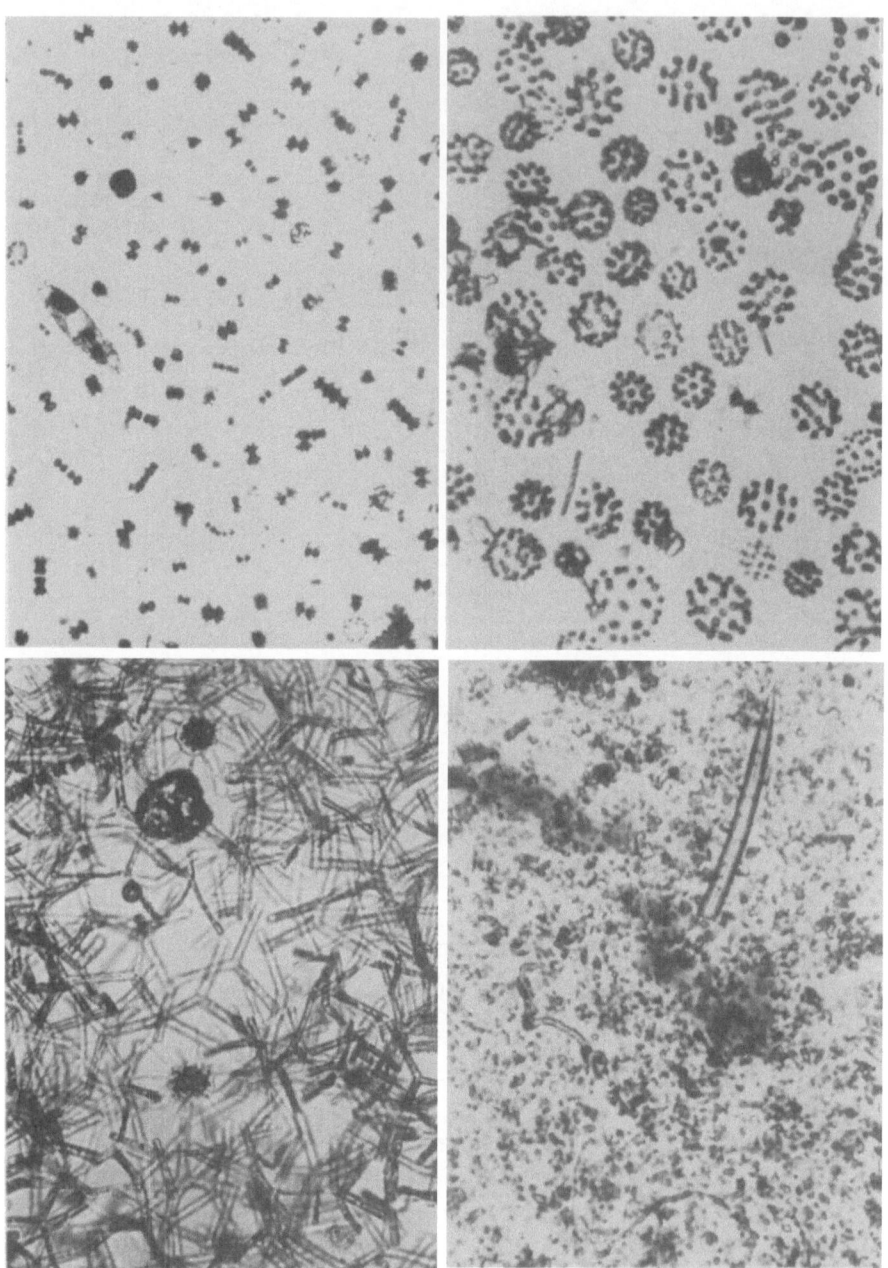

Pl. X. Plankton: a. Lake Brokopondo at Afobaka, 19.IV.1968. (× 75). — b. Lake
Brokopondo at Afobaka, 4.III.1964 (× 90). — c. Suriname River at Afobaka,
27.XI.1963 (× 180). Suriname River at Paranam, 16.XI.1963 (× 60)

PLATE XI

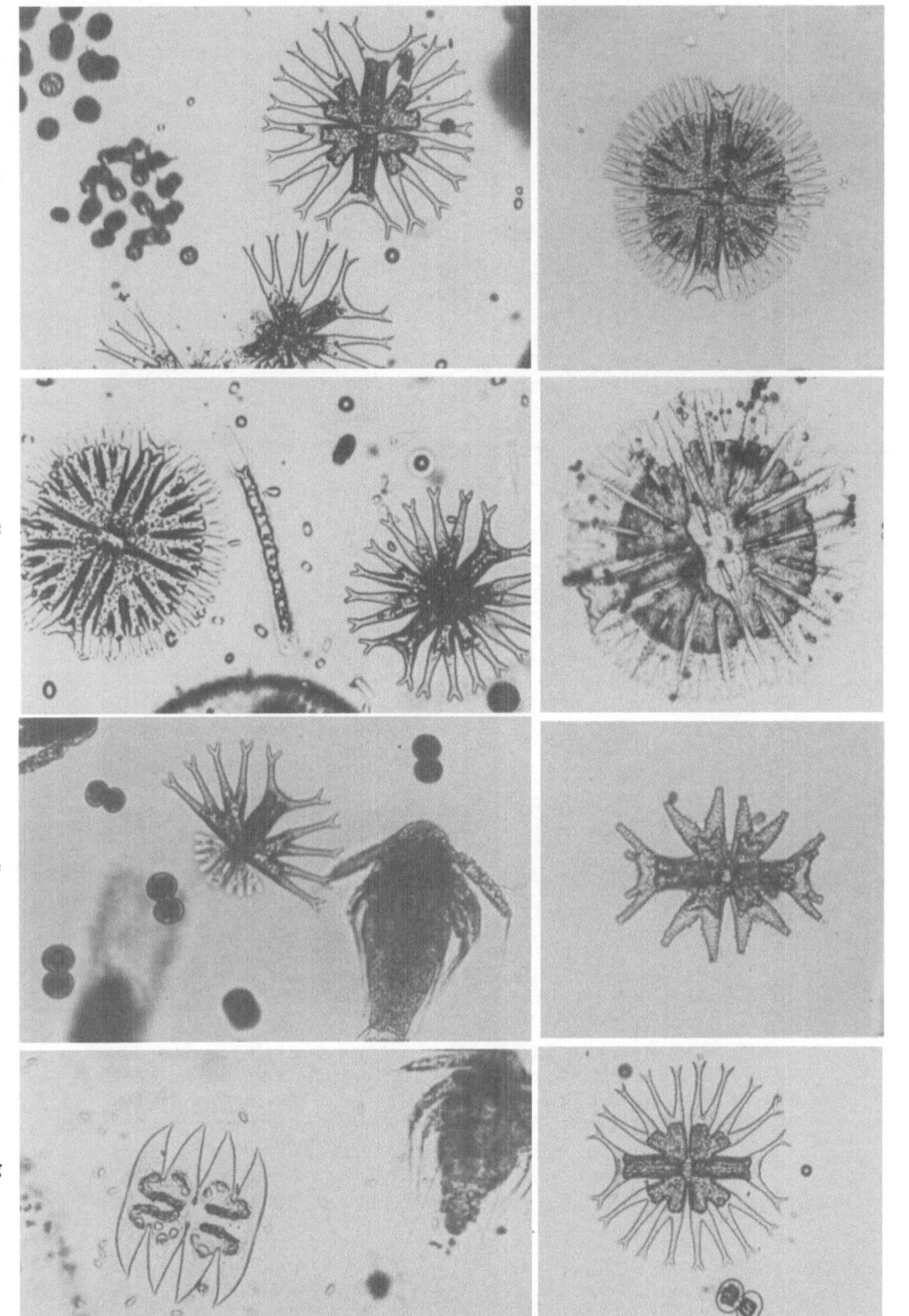

Pl. XI. Desmidiaceae (abt. × 100): a. *Micrasterias radiata* var. *brasiliensis*, Afobaka 29.III.1967. — b. *M. radiata*, Sara 3.IV.1967. — c. *M.* sp., Afobaka, 29.III.1967. — d. *M. schweinfurthii*, Beerdotti 8.III.1967. — e. *M. radiata*, Sara 4.I.1967. — f. *M. mahabuleshwarensis*, Kabel railroad 19.V.1967. — g. *M. laticeps*, Lokus 22.III.1967. — h. *M. radiata* var. *brasiliensis*, Afobaka 26.IV.1967.

PLATE XII

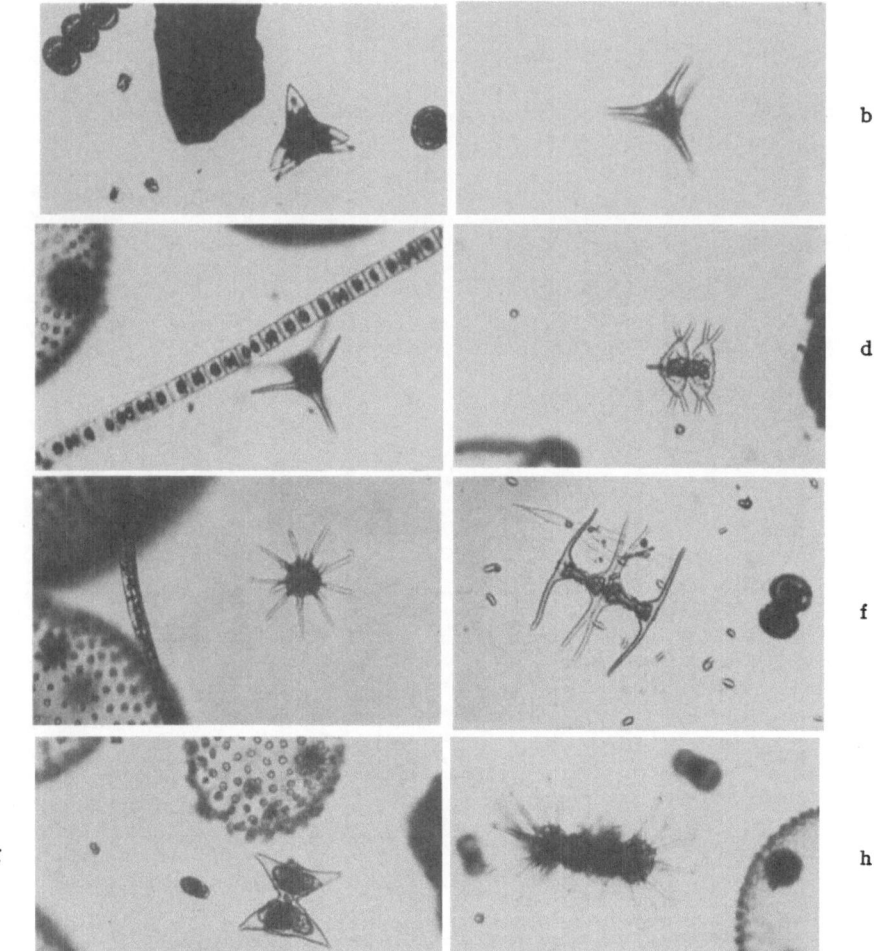

Pl. XII. Desmidiaceae (abt. × 100): a. *Staurastrum subgrande*, Beerdotti 5.V.1967. — b. *S. sebaldi*, Grankreek 14.IV.1967. — c. *S. sebaldi*, Beerdotti 22.III. 1967. — d. *S. setigerum*, Grankreek 4.IV.1967. — e. *S. leptacanthum*, Grankreek 7.II.1967. — f. *S. leptocladum*, Afobaka 29.III.1967. — g. *S. curvatum*, 6.IV.1967. — h. *S. leptacanthum*, Afobaka 24.V.1967.

PLATE XIII

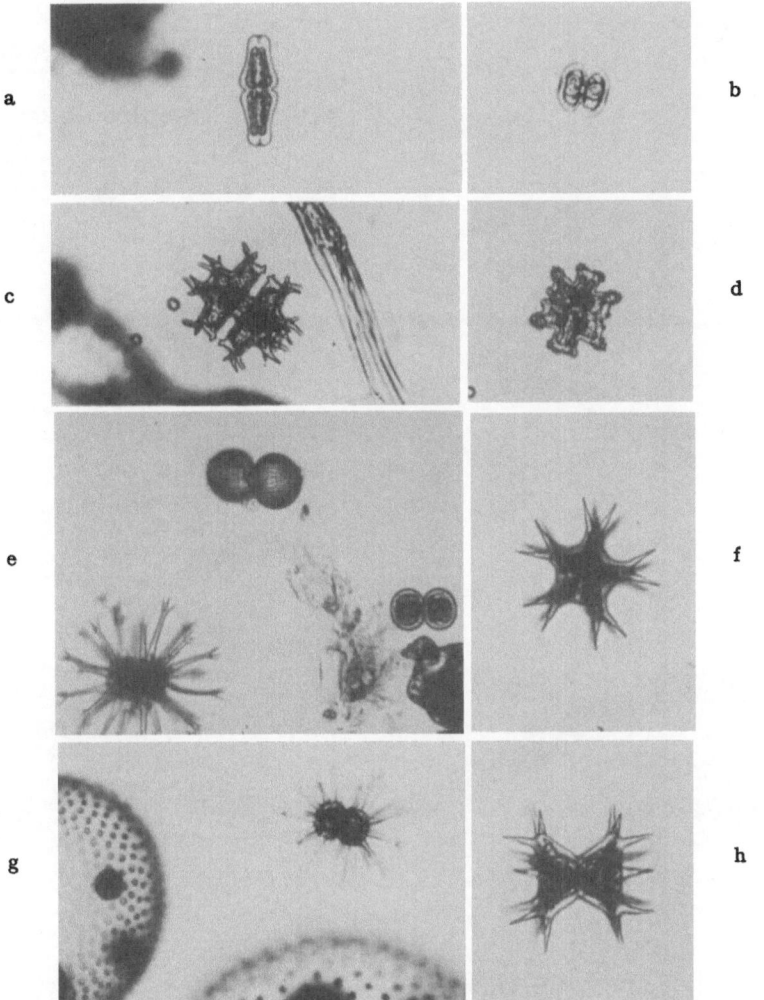

Pl. XIII. Desmidiaceae (abt. × 100): a. *Tetmemorus* sp., Nickerie River 4.IV.1967. — b. *Arthrodesmus convergens*, Grankreek 7.II.1967. — c. *Xanthidium trilobum*, Beerdotti 15.II.1967. — d. *Euastrum spinulosum*, Pittipratti 4.IV.1967. — e. *Cosmarium* sp. 7, Sara 31.V.1967. — f. *Staurastrum sinuatum*, Pittipratti 4.IV.1967. — g. *S. sinuatum*, Pittipratti 4.IV.1967. — h. *S. leptacanthum*, Afobaka 24.V.1967.

PLATE XIV

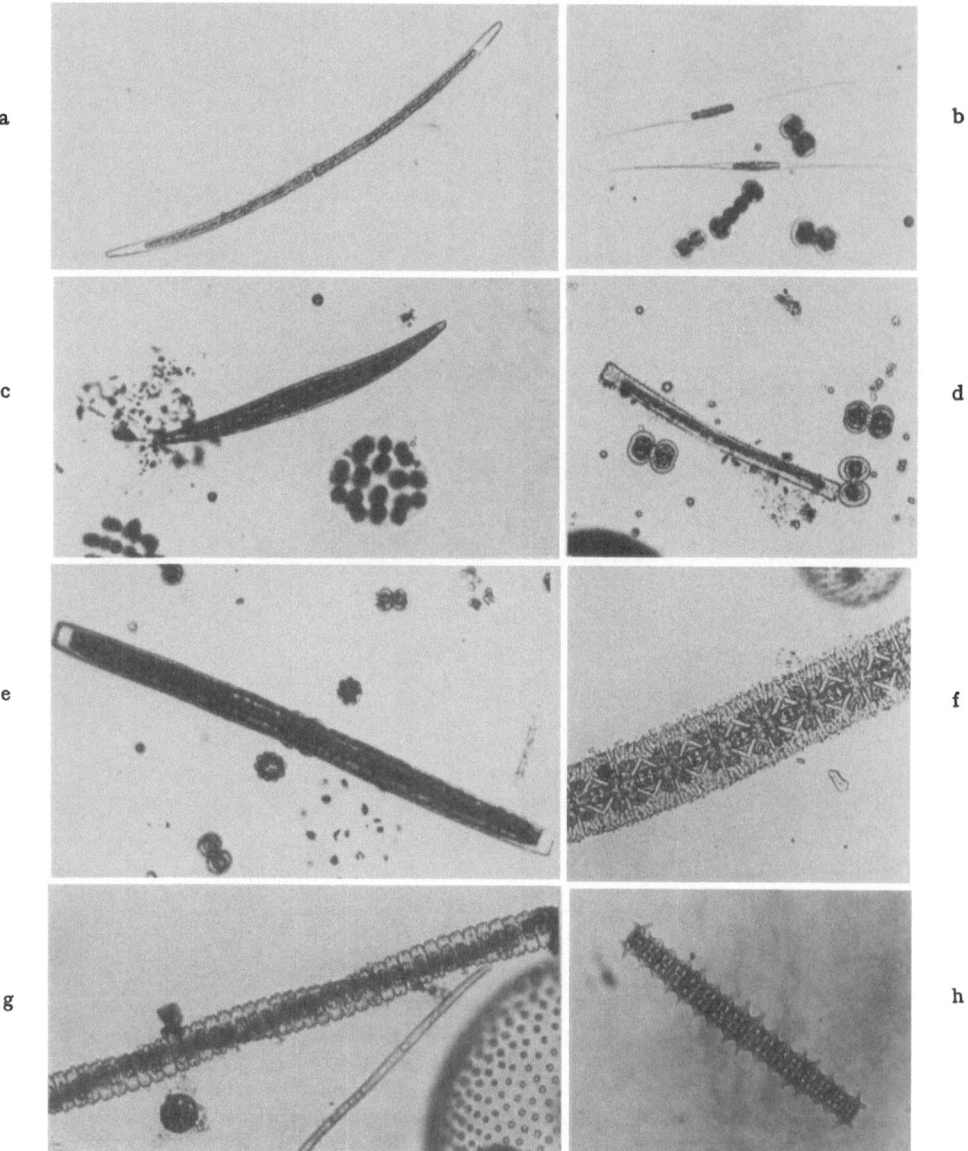

Pl. XIV. Desmidiaceae (abt. 75): a. *Closterium* sp., Grankreek 13.IV.1967.
— b. *C. kuetzingii*, Afobaka 26.IV.1967. — c. *C.* sp., Afobaka 18.I.1967. — d. *Gonatozygon* sp., Lokus 16.III.1967. — e. *Pleurotaenium* sp., Afobaka 11.I.1967. —
f. *Micrasterias foliacea*, Beerdotti 18.I.1967. — g. *Desmidium swartzii*, Kabel railroad
19.V.1967. — h. *Onychonema* sp., Beerdotti 18.I.1967.

PLATE XV

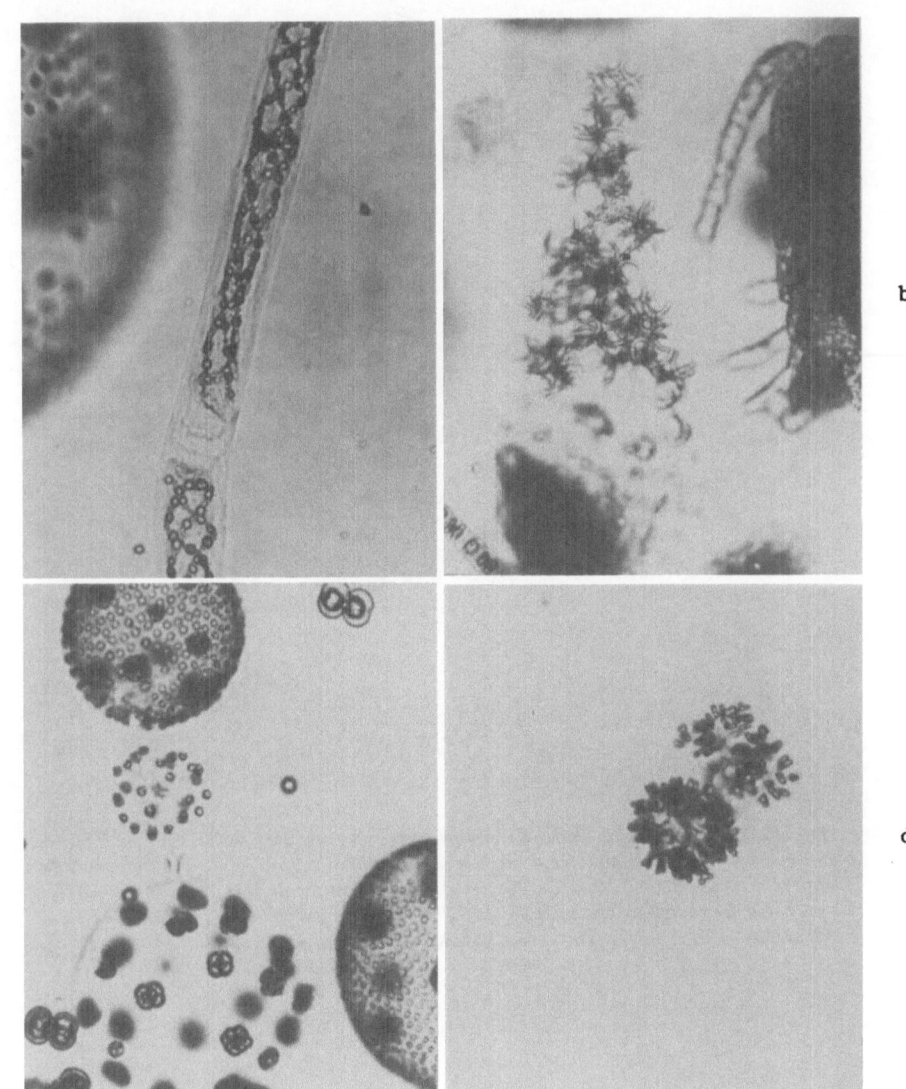

Pl. XV. Chlorophyceae (abt. × 75): a. *Spirogyra* sp., Beerdotti 1.III.1967. —
b. *Selenastrum* sp., Kabel railroad 19.V.1967. — c. *Dictyosphaerium* sp., Pittipratti
4.IV.1967. — d. *Dimorphococcus* sp., Afobaka 26.IV.1967.

PLATE XVI

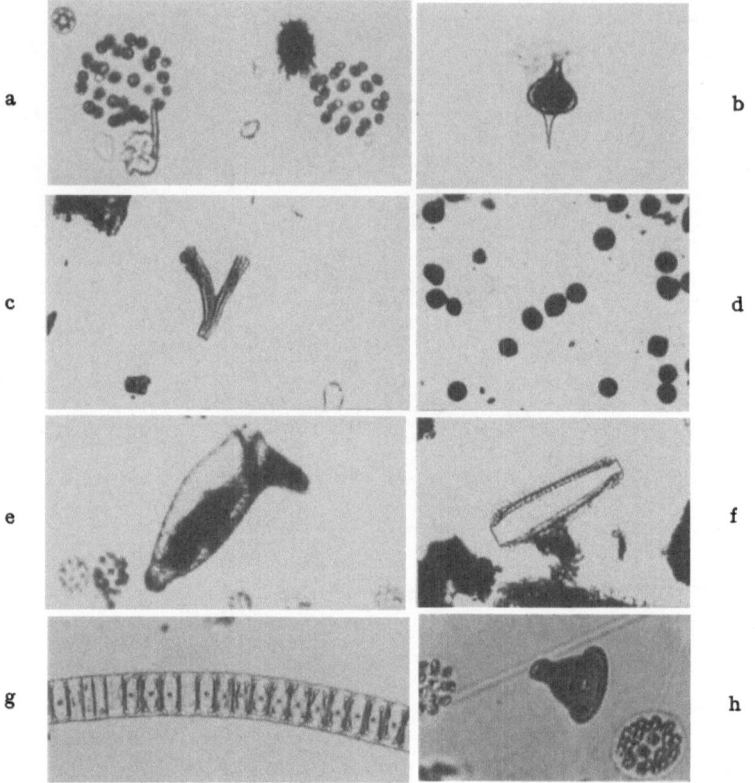

Pl. XVI. Miscellaneous: a. *Trachelomonas armatus*, Sara 23.II.1967 (× 75). —
b. *Strombomonas ensifera*, Sara Kreek 23.II.1967 (× 75). — c. *Rhipidodendron
huxleyi*, Pokigron 5.IV.1967 (× 75). — d. Saprobic unicellulars, Grankreek 13.IV.
1967. — e. Bryozoa-zoecium, 23.II.1967 (× 30). — f. *Surirella* sp., Suriname River,
Mamadam, 3.IV.1967 (× 75). — g. Diatom. — h. Rhizopod, Lokus 19.II.1964 (× 3).

PLATE XVII

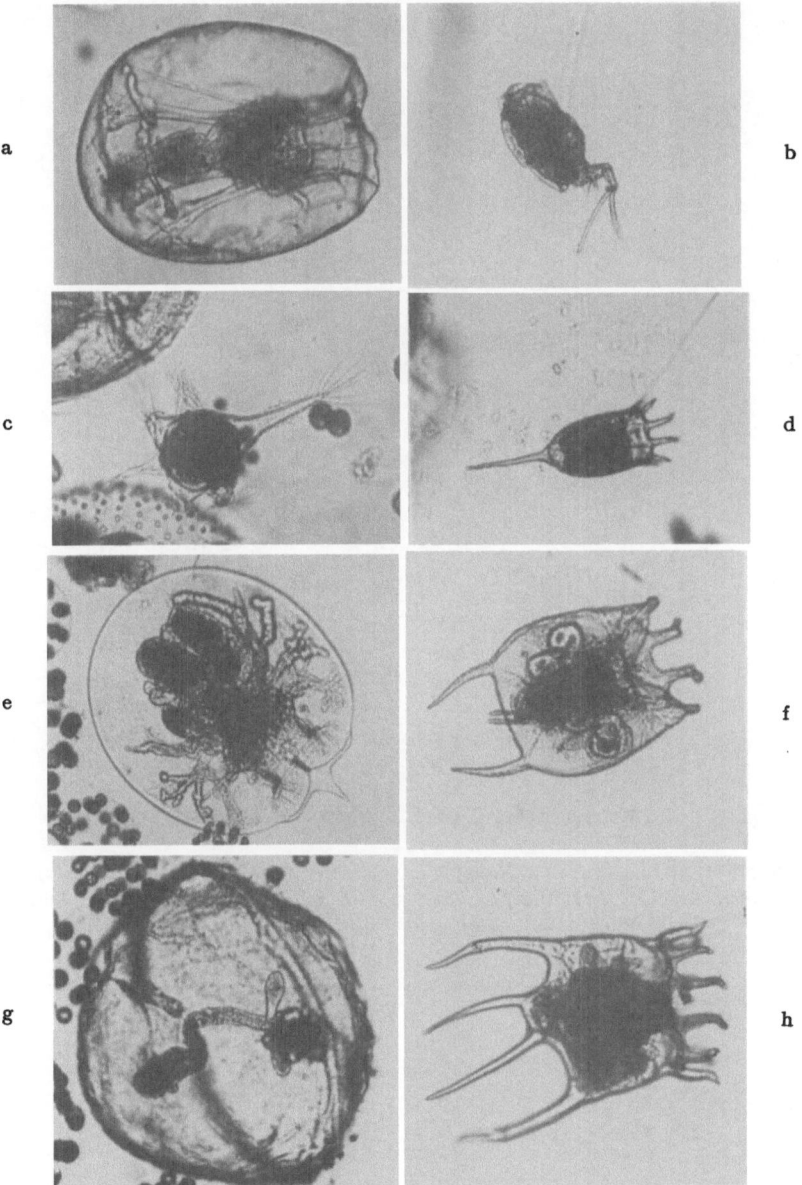

Pl. XVII. Rotifera: a. *Asplanchna* sp., Grankreek 13.IV.1967 (× 75). — b. *Keratella americana*, Koenkoen 3.IV.1967 (× 75). — c. *Hexarthra mira*, Afobaka 4.I.1967 (× 75). — d. *Trichotria tetractis*, 23.II.1967 (× 75). — e. *Testudinella mucronata*, Sara Kreek 23.II.1967 (× 30). — f. *Brachionus quadridentata*, Adawai 3.IV.1967 (× 30). — g. *Trochosphaera meridionalis*, Afobaka 25.I.1967 (× 75). — h. *Platyias patulus*, Koenkoen 3.IV.1967.

PLATE XVIII

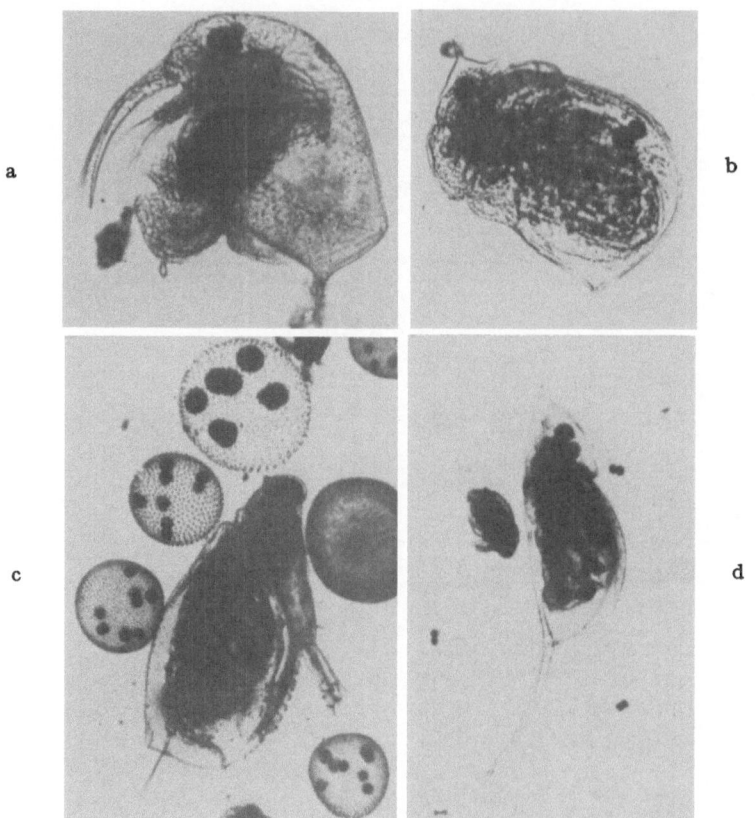

Pl. XVIII. Cladocera: a. *Bosmina* sp., Sara 3.V.1967 (× 75). — b. *Ceriodaphnia cornuta*, Afobaka 29.II.1967 (× 75). — c. *Diaphanosoma brachyurum*, Afobaka 29.III.1967 (× 30). — d. *Daphnia* sp., Afobaka 4.I.1967 (× 30).

FRESHWATER SPONGES OF SURINAME

by

INÉS EZCURRA DE DRAGO

(Instituto Nacional de Limnología, Santo Tomé (Santa Fé), Argentina)

This paper is the first contribution to the knowledge of the fresh-water sponges of Suriname. Four species have been identified up till now: *Metania spinata* (Carter, 1881), *Trochospongilla paulula* (Bo-werbank, 1863), *Radiospongilla crateriformis* (Potts, 1882), and *Drulia uruguayensis* Bonetto & Ezcurra de Drago, 1969. Since in most instances slides were the only available source of data, details on external structure cannot be supplied.

The author is indebted to Dr. D. C. GEIJSKES of the Rijksmuseum van Natuurlijke Historie, Leiden, who kindly provided a great deal of the material for study, as well as to Dr. W. VERVOORT, Leiden, and to Drs. P. LEENTVAAR, R.I.N., Leersum, who also sent specimens. The author also acknowledges the authorities and members of the staff of the Invertebrates Department of the British Museum. Special thanks are due to Miss SHIRLEY STONE who facilitated the study of South American material, and to Prof. J. BENOIT, Invertebrates Department of the Musée Royal de l'Afrique Centrale, Tervuren, who provided African material of great interest.

The author is Fellow member of the Research Career of the Consejo Nacional de Investigaciones Científicas y Técnicas.

The material studied will be deposited in the Rijksmuseum van Natuurlijke Historie, Leiden.

Metania spinata (Carter, 1881)   Fig. 41

Tubella spinata CARTER, 1881; TRAXLER, 1895; WELTNER, 1895; GEE, 1931; GEE, 1931; Gee, 1933; PENNEY, 1960.
Metania spinata, PENNEY & RACEK, 1968.

Megascleres smooth, fusiform, slender and slightly curved amphioxea. — Microscleres fusiform, slender and spined amphioxea, with largest central spines. Very numerous in all parts of the symplasm. — Gemmules large, ovoid. Pneumatic layer thick, with polygonal air spaces. Gemmoscleres embedded in this layer radially,

176

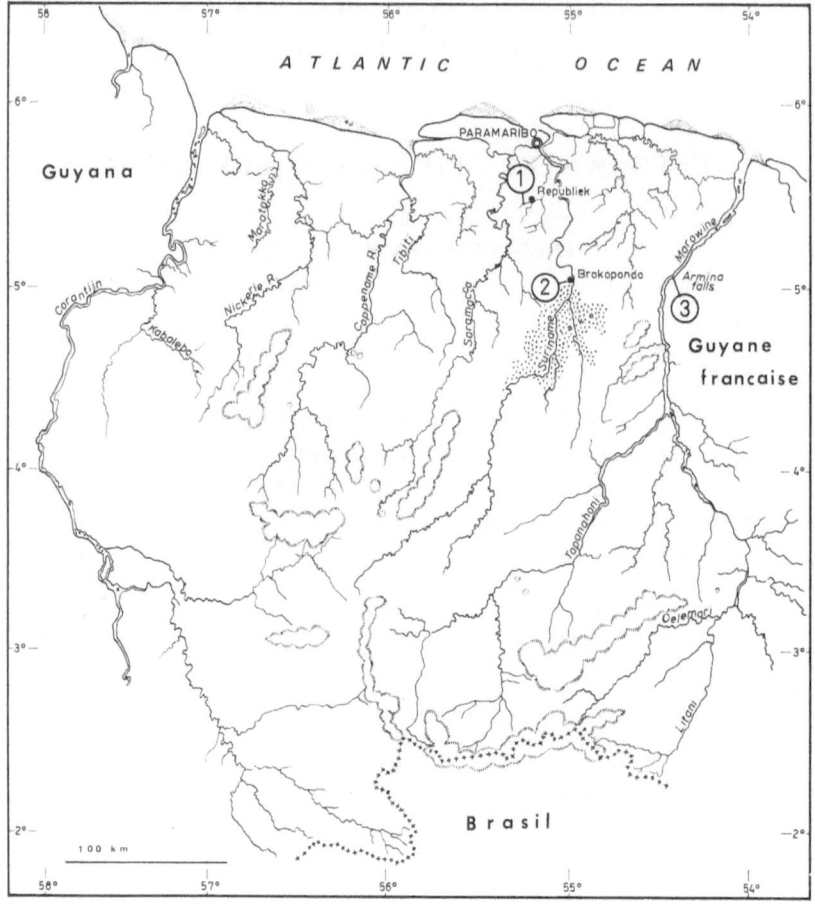

Fig. 40. Sketch map of SURINAME with localities. – 1: Swamp near Para river, tributary of the Suriname river. 2: Brokopondo, north of the man-made Brokopondo Lake. 3: Armina falls, Marowijne river.

projecting their terminal ends beyond the surface of pneumatic coat. Foramen tubular, foraminal tube straight, surrounded by slanting gemmoscleres. — Around the gemmules there are megascleres feebly curved to almost straight, ci'indrical amphioxea, abruptly sharpened on the tips, throughout their length covered with conspicuous spines. — Gemmoscleres tubelliform, with the lower rotule polygonal or irregularly circular, with slightly recurved and entire margin.

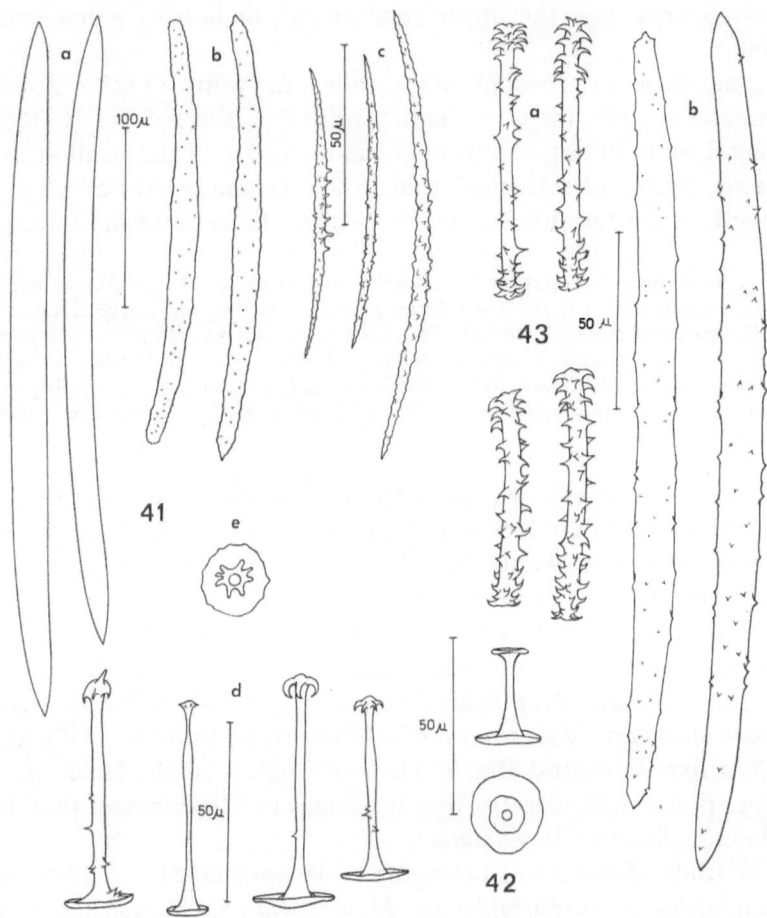

Fig. 41. *Metania spinata* (Carter). – a: Smooth megascleres. b: Rough megascleres. c: Microscleres. d: Gemmoscleres. e: Detail of lower rotule of gemmosclere.

Fig. 42. *Trochospongilla paulula* (Bowerbank). – Gemmosclere with detail of rotule.

Fig. 43. *Radiospongilla crateriformis* (Potts). – a: Gemmoscleres. b: Megascleres.

Radial striations of this rotule often well developed. Upper rotule invariably knoblike, bearing a varying number of large, recurved marginal spines or represented by a rounded knob with some little spines. Shaft proportionally long and thin, generally with a little enlargement near the upper end, smooth or bearing a few small spines.

Smooth megascleres 300–390 μ in length; width 17–22 μ. Spined megascleres 115–140 μ in length; width 7 μ. Microscleres 70–105 μ long. Length of largest axis of gemmules 850 μ, of the smallest axis 540 μ; length of foraminal tube 100 μ. Gemmoscleres 50–62 μ in length; diameter of lower rotule 15–25 μ; diameter of shaft 3–5 μ.

Collected by D. C. GEIJSKES from floating stem in swamp of the Coropina Kreek, Para river, tributary of the lower Suriname river, near Republiek, July 1955.

Comparative material studied: *Metania vesparioides* (Annandale, 1908, Great Pond at Mudon, near Moulmeer, Amherst District, Burma. *Metania reticulata* (Bowerbank, 1863), Amazon river. *Metania mello-leitao* (Machado, 1945), Tapirapé river, Brazil. *Metania lissostrongyla* (Burton, 1938) (now *M. vesparia*), lake Tumba, Congo.

Up till now the knowledge of *Metania spinata* (Carter) was limited to its original description and to TRAXLER's work, dealing with material from San Pablo, Brazil. It was originally described as *Tubella spinata* Carter and then transferred to *Metania* Gray by PENNEY & RACEK (1968), when redefining the genus. They include in. *Metania* only two South American species: *M. reticulata* (Bowerbank, 1863), and *M. spinata* (Carter, 1881), excluding *Tubella melloleitao* Machado, 1945, from the Tapirapé river, tributary of the Araguaia river in central Brazil. The examination of the Holotype of this species demonstrates that it belongs to *Metania*, and that it is closely related to *M. reticulata*.

A study of specimens belonging to *M. spinata*, *M. reticulata*, *M. mello-leitao*, *M. vesparioides* and *M. vesparia* (von Martens), demonstrated the validity of *M. spinata*. As a matter of fact, this species can be clearly differentiated from the other South American species ot the genus (*M. reticulata* and *M. mello-leitao*) because of its gemmoscleres (much longer, with an almost smooth axis), as well as its megascleres (much longer and thinner). It seems more closely related to *M. vesparioides*, because of the megascleres and gemmoscleres structure, though the latter are much smoother in *M. spinata*.

Unfortunately, since we have hardly any material of this last species (mainly gemmules), we cannot say more about its relation to *M. vesparioides*. A comparative study between these two species and

M. innominata Burton, 1938, would also be of great interest. They apparently constitute a quite homogenous group, represented in South America, Africa and India. The difference with the rest of the *Metania* species would be the conformation of the megascleres. We cannot agree, so far, with PENNEY & RACEK (1968) in considering *M. vesparioides* as a probable *M. vesparia* subspecies.

Trochospongilla paulula (Bowerbank, 1863)   Fig. 42

Spongilla paulula BOWERBANK, 1863.
Tubella paulula, BONETTO & EZCURRA, 1962; BONETTO & EZCURRA, 1964; BONETTO & EZCURRA, 1967.
Trochospongilla paulula, PENNEY & RACEK, 1968 (and synonymy); VOLKMER-RIBEIRO & ROSA-BARBOSA, 1972; BONETTO & EZCURRA, 1973.
Trochospongilla latouchiana ANNANDALE, 1907; PENNEY & RACEK, 1968 (and synonymy).

Megascleres slender, feebly curved, smooth, fusiform amphioxea. — Microscleres absent. — Gemmules spherical, pneumatic layer well developed, granular, foramen tubular with a short tube. — Gemmoscleres embedded radially, birrotulates with a slender and smooth shaft. Rotules irregularly circular and of unequal diameter, the upper one considerably smaller than the lower, both recurved in the same direction.

Megascleres 230–250 μ in length; width 8–12 μ. Diameter of gemmules 250–270 μ. Gemmoscleres 24–27 μ in length; diameter of lower rotule 18–25 μ, of upper rotule 8–12 μ; diameter of shaft 3–4 μ.

Collected by D. C. GEIJSKES from floating stem in swamp of the Coropina Kreek, Para river, near Republiek, July 1955.

Trochospongilla paulula (Bowerbank) is widely spread in South America. It has been found in tributaries of the Amazonas (VOLKMER-RIBEIRO & ROSA BARBOSA, 1972), in the upper and middle reaches of the Paraná river (BONETTO & EZCURRA, 1962, 1964, 1967, 1973) and in central Uruguay river (BONETTO & EZCURRA, 1967). A thorough study on the variability of the species has recently been performed by BONETTO & EZCURRA (1973). They demonstrate the necessity of synonymizing *Trochospongilla latouchiana* Annandale. In this way, *T. paulula* represents – among the Porifera, and according to the known literature – the first species with a continuous distribution from Australia to India and China, Africa and South America.

Radiospongilla crateriformis (Potts, 1882)   Fig. 43

Meyenia crateriformis POTTS, 1882.
Ephydatia crateriformis var. *arndti* CARVALHO, 1942.
Ephydatia crateriformis paranensis BONETTO & EZCURRA, 1964b.
Radiospongilla crateriformis, PENNEY & RACEK, 1968 (and synonymy).

Megascleres fusiform, abruptly pointed amphioxea, sparsely and completely microspined, except at tips. — Microscleres absent. — Gemmoscleres slender amphistrongyla, with a varying number, generally abundant, of straight or recurved conical spines, which form several radiating rows at the ends of the spicules.

Megascleres 205–250 µ in length; width 8–11 µ. Gemmoscleres 55–75 µ in length; width 3–6 µ.

Collected by P. LEENTVAAR near the dam of lake Brokopondo, Suriname river, 1964.

Ephydatia crateriformis (Potts) was included in *Radiospongilla* Penney & Racek. The presence of this species in South America has already been pointed out by CARVALHO (1942) for the São Paulo State, as well as for the middle Paraná river, in Santa Fe (BONETTO & EZCURRA, 1964). CARVALHO describes a new form, *E. crateriformis* var. *arndti* which was not recognized by PENNEY & RACEK (1968). BONETTO & EZCURRA distinguish a subspecies *paranensis*, based on the differences shown by the Paraná river material.

The Suriname material presents intermediate characteristics between the form described by CARVALHO and subsp. *paranensis*. In fact, the conformation of the gemmoscleres is quite the same as in *E. crateriformis paranensis*; on the other hand, the megascleres are rather similar to the "tornotas" type of CARVALHO, since they have abruptly finished tips, instead of the gradually sharpened ones as are typical in subsp. *paranensis*. For this reason I do not consider it wise to sustain the validity of the subspecies from the Paraná river.

Drulia uruguayensis Bonetto & Ezcurra, 1969 Fig. 44

Drulia uruguayensis BONETTO & EZCURRA, 1969; Bonetto & Ezcurra, 1970.

Megascleres stout, smooth, slightly curved and fusiform amphioxea with pointed tips. — Microscleres scarce, thin and pointed amphioxea, slightly curved, microspined in their tips and with larger

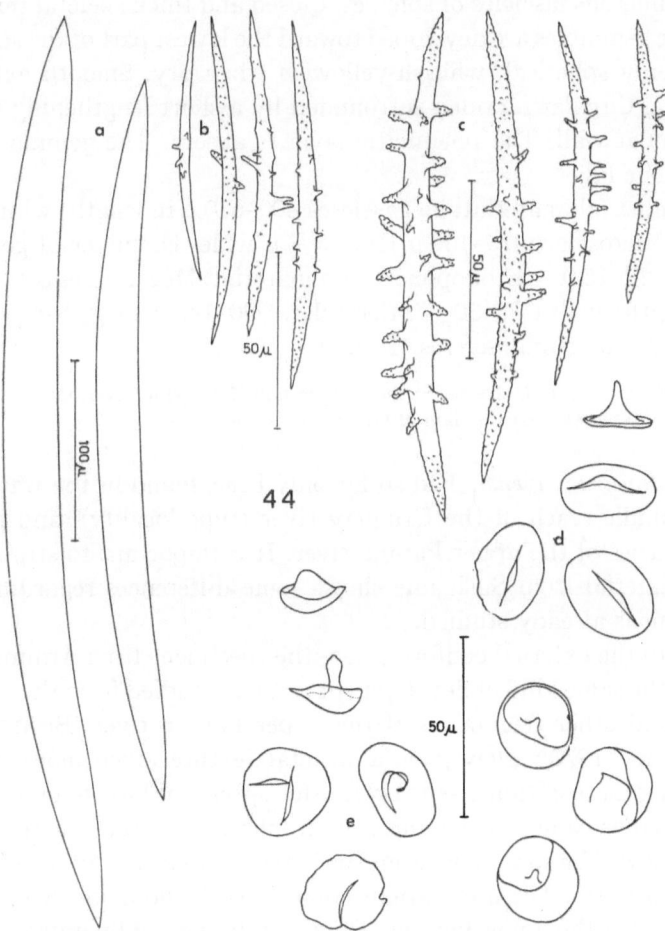

44

Fig. 44. *Drulia uruguayensis* Bonetto & Ezcurra. – a: Megascleres, material from the Armina falls. b: Microscleres, from the Armina falls. c: Microscleres, from the Brokopondo dam. d: Gemmoscleres, material from the Armina falls. e–f: Gemmoscleres from the Brokopondo dam.

microspined projection in the central part. — Gemmoscleres par-
muliform, with circular or irregular margins, conical or longitudinal-
ly umbonated, some of them very modified, with the rims variably
curved toward the upper and inner parts of the spicules.

It is a sponge forming incrustations; its colour, when dry, is light
hazel. Oscula inconspicuous, hispid surface, due to the projection of
the numerous fascicles of spicules. Closed and thick skeletal texture.
— The gemmules are developed toward the lowest part of the sponge.
Unevenly spherical, whitish-yellowish when dry. Smooth external
surface. Circular foramen surrounded by a short lengthening of the
gemmular wall. The pneumatic layer is absent. The gemmoscleres
lie in one layer.

Armina fall material: Megascleres 400–670 μ in length; width 35–
70 μ. Microscleres 75–100 μ long, 5–6 μ wide. Diameter of gemmoscleres 12–15 μ. – Brokopondo dam material: Megascleres 370–480 μ
in length; width 25–30 μ. Microscleres 80–140 μ long, 5–6 μ wide.
Diameter of gemmoscleres 12–25 μ.

Collected by D. C. GEIJSKES in the Armina falls, Marowijne river, Oct. 1952, and
by P. LEENTVAAR near the dam of lake Brokopondo, 1964.

Drulia uruguayensis had so far only been found in the waters of
the middle reach of the Uruguay river (type locality), and in the
tributaries of the upper Paraná river. It is important to stress that
the material from Suriname shows some differences regarding the
specimens already studied.

As to the external conformation, the specimens from Armina falls
show the same kind of development as those studied from the Iguazú
falls and other tributaries of the upper Paraná river (BONETTO &
EZCURRA, 1970). They gave a skeletal texture much more closed,
forming incrustations, as is typical for species of lotic habitats. The
material has much bigger and stouter microscleres than those known
until now. The specimens from Brokopondo lake have very modified
gemmoscleres. These variations coincide with those observed in the
material of the upper Paraná. They may be caused by unfavourable
environment and be of little taxonomic value for the moment.

183

REFERENCES

ANNANDALE, N., 1908. Notes of freshwater sponges IX. Preliminary notice of a collection from Burma, with the description of a new species of Tubella. *Rec. Ind. Mus. 2*: 157–158.

BONETTO, A. A. & EZCURRA DE DRAGO, I., 1962. Dos nuevas esponjas para el Paraná medio. *Physis 23*: 209–213.

BONETTO, A. A. & EZCURRA DE DRAGO, I., 1964a. Esponjas del Paraná medio e inferior. *Anais II Congr. Lat. Amer. Zoologia, São Paulo I*: 245–260.

BONETTO, A. A. & EZCURRA DE DRAGO, I., 1964b. Nuevas esponjas de agua dulce de la República Argentina. *Physis 24* (68): 329–336.

BONETTO, A. A. & EZCURRA DE DRAGO, I., 1967. Esponjas del noreste argentino. *Acta Zool. Lill. 23*: 331–348.

BONETTO, A. A. & EZCURRA DE DRAGO, I., 1969. El género Drulia Gray en el río Uruguay (Porifera, Spongillidae). *Physis 28* (76) 1968: 211–216.

BONETTO, A. A. & EZCURRA DE DRAGO, I., 1970. Esponjas de los afluentes del alto Paraná en la provincia de Misiones. *Acta Zool. Lill. 27*: 37–61.

BONETTO, A. A. & EZCURRA DE DRAGO, I., 1973. Las esponjas del género Trochospongilla Vejdovsky en aguas argentinas. *Physis (B) 32* (84): 13–18.

BOWERBANK, J. S., 1863. A monograph of the Spongillidae. *Proc. Zool. Soc. London:* 440–472.

CARTER, H. J., 1881. History and classification of the known species of Spongilla. *Ann. Mag. Nat. Hist. 7*: 247–250.

CARVALHO, J. P., 1942. Ocorrencia de Ephydatia crateriformis (Potts) na America do Sul. *Bol. Faculdade Filosofia Cienc. Let. 15*: 267–279.

GEE, N. G., 1931. A contribution toward an alphabetical list of the known freshwater sponges. *Peking Nat. Hist. Bull. 5*: 31–52.

GEE, N. G., 1933. Fresh-water sponges, genus Tubella. *Peking Nat. Hist. Bull. 7*: 237–252.

GRAY, M. E., 1867. Notes on the arrangement of sponges with description of some new genera. *Proc. Zool. Soc. London*: 492–558.

MACHADO, O. X. DE, 1948. Contribution à l'étude de la faune du Brasil. Nouvelle espèce de spongiaire fluvial: Tubella mello-leitaoi O. Machado. *Bull. Soc. Zool. France 72*: 133–135.

PENNEY, J. T., 1960. Distribution and bibliography (1892–1957) of the freshwater sponges. *Univ. S. Carolina Publ. (3) 3*: 1–97.

PENNEY, J. T. & RACEK, A. A., 1968. Comprehensive revision of a worldwide collection of freshwater sponges (Porifera, Spongillidae). *Bull. U.S. Nat. Mus. 272*: 1–184.

POTTS, E., 1882. Three more freshwater sponges. *Proc. Acad. Nat. Sci. Philad. 34*: 12–14.

POTTS, E., 1887. Contributions toward a synopsis of the American forms of freshwater sponges. *Proc. Acad. Nat. Sci. Philad. 39*: 158–279.

TRAXLER, L., 1895. Die Schwammspikule des Schlammes im See Héviz. *Földtani Közlony 25*: 142–145.

VOLKMER-RIBEIRO, C. & ROSA-BARBOSA, R. DE, 1972. On Acalle recurvata (Bowerbank, 1863) and an associated fauna of other freshwater sponges. *Rev. Brasil. Biol. 32* (3): 303–317.

WELTNER, W., 1895. Spongillidenstudien III. Katalog und Verbreitung der bekannten Süsswasserschwämme. *Archiv Naturgesch. 61*: 114–144.

PARASITIC MITES OF SURINAM
XXXIII. Feather mites (Analgoidea)

by
V. ČERNÝ
(Institute of Parasitology, Czechoslovak Academy of Sciences, Prague)

and
F. S. LUKOSCHUS
(Zoölogisch Laboratorium, Katholieke Universiteit, Nijmegen)

The mites listed in the present paper have been collected from July to October 1971 by the junior author and Drs. N. J. J. Kok during their stay in Surinam with financial aid of the Netherlands Foundation for the Advancement of Tropical Research (Wotro). Mites have been collected occasionally mainly from birds, found dead on the roads, and from captured birds which died in the pet shop "Tropical Wildlife", Paramaribo. As in many instances the hosts were decayed, few mites could be prepared. The relation of formerly described species to new species clearly shows our poor knowledge of parasites on birds from this region.

Most of the species have been described by the senior author (Černý, 1974a, b, 1975). This paper contains descriptions of new *Trouessartia* species and a survey of all species found. The taxa are arranged in alphabetical order.

In this survey are not included: 1) a trouessartiin genus near *Calcealges* from *Amazilia fimbriata*, represented only by nymphs; 2) a new genus representing a new subfamily of syringicolous mites (fide dr. Gaud); 3) one female from *Columbigallina talpacoti* belonging to a new species of a new genus near *Diplaegidia* (to be described by dr. Gaud).

The authors are deeply indebted to Dr. J. Gaud (Rennes) for critical remarks concerning the systematic position of some species.

Family ALLOPTIDAE Gaud & Mouchet, 1958

Subfamily Trouessartiinae Gaud & Mouchet, 1958

Genus **Trouessartia** Canestrini, 1899

Trouessartia aedon sp. n.   Fig. 45-46

Host and locality. – House wren, *Troglodytes aedon* Vieillot, 1808, local name gadofawroe, Paramaribo, 18.VIII.1971 (holotype ♂ and 4 ♂♂, 3 ♀♀, 5 nymphs, 1 larva).

Deposition of type material. – Rijksmuseum van Natuurlijke Historie, Leiden; National Collection of Surinam, Paramaribo; University of Georgia, Athens; Institute of Parasitology, Prague.

Male (holotype) (Fig. 45): Total length (excl. lamellae) 476 μ, idiosomal length 437 μ, width 228 μ. Propodosomal shield 134 × 138 μ, distant from scapular shields: *sce-sce* 93 μ. Setae *l I* 35 μ. Hysterosomal shield almost straight anteriorly, slightly biconcave

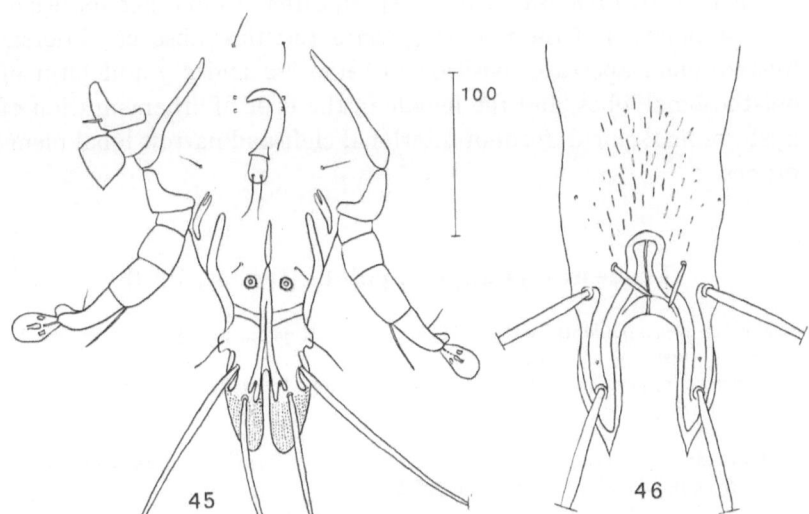

Fig. 45. *Trouessartia aedon* sp. n. – Male, body terminus, ventral view.
Fig. 46. *Trouessartia aedon* sp. n. – Female, body terminus, dorsal view.

laterally. Terminal lamellae elongate with straight margins. Sterno-ventral sclerites absent. Setae *sh* 22 μ, lanceolate, setae *sR* 19 μ. Genital organ 43 μ. Setae *c 2* on weakly sclerotized subgenital shield, close together (6 μ) and slightly posterior to setae *c 3*. Epimerites IVa of about the same length as adanal apodemes. Adanal discs 13 × 13 μ, in mid-distance between caudal part of genital organ and body terminus. Translobar apodeme developed.

Female (allotype) (Fig. 46): Total length 560 μ, idiosomal length 515 μ, width 231 μ. Propodosomal shield 145 × 154 μ, distant from scapular shields: *sce-sce* 110 μ. Setae *l 1* 33 μ. Hysterosomal shield almost straight anteriorly and slightly concave anterolateral-ly, with fissiform lacunae between setae *d 3* and *d 4*. Supranal con-cavity present, connected with interlobal cleft. Setae *d 4* 33 μ, lan-ceolate. Opisthosomal lobes slightly converging posteriorly, with only narrow external and internal membrane terminating in acute tip. Interlobal cleft broad, with large transverse membrane on its bottom perforated by the spermathecal duct. Setae *sh* 24 μ, lanceo-late, setae *sR* 16 μ. Bases of setae *c 1* contiguous with pregenital apodeme, *c 1-c 2* 16 μ.

The male of *Trouessartia aedon* sp. n. differs from other species in the combination of the following characteristics: absence of dorsal hysterosomal aperture, position of setae *c 2* and *c 3* and form of opisthosomal lobes, and the female in the type of ornamentation of hysterosomal shield, form of interlobal cleft and narrow lobal mem-branes.

Trouessartia appendiculata (Berlese, 1884)

Pterocolus appendiculatus Berlese, 1884. A.M.S., Repert., ser. *5*, nr *27* & A.M.S.,
    fasc. *24*, nr. *7* (descr. ♂♀).
Trouessartia appendiculata, Canestrini & Kramer, 1899. Tierreich *7*: 121 (short
    diagn. ♂♀).

Host and locality. – Black-collared swallow, *Atticora melanoleuca* (Wied, 1820), Weg naar Zee, 10.IX.1971 (1 ♂, 2 ♀♀).

The species is known from various European and African Hirundinidae (GAUD & TILL, 1961). It is reported for the first time from South America.

187

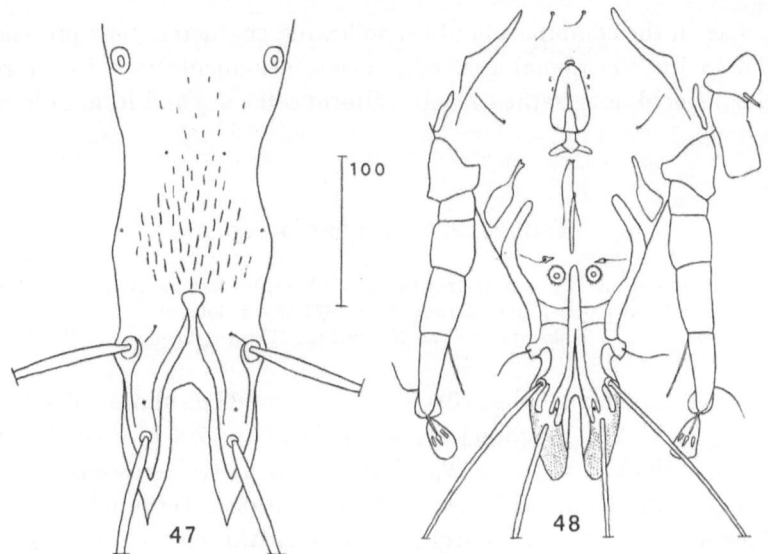

Fig. 47. *Trouessartia fissispina* sp. n. – Female, body terminus, dorsal view.

Fig. 48. *Trouessartia megaplax* sp. n. – Male, body terminus, ventral view.

Trouessartia fissispina sp. n.   Fig. 47

Host and locality. – Yellow-bellied elaenia, *Elaenia flavogaster* (Thunberg, 1822), Weg naar Zee, 10.IX.1971 (1 ♀ holotype).
Deposition. – Rijksmuseum van Natuurlijke Historie, Leiden.

Female (holotype) (Fig. 47): Total length 547 μ, idiosomal length 500 μ, width 203 μ. Propodosomal shield 138 × 137 μ, distant from scapular shields: *sce-sce* 95 μ. Setae *l 1* 41 μ. Hysterosomal shield almost straight anteriorly, with large apertures and fissiform lacunae between hysterosomal apertures and supranal concavity which is subtriangular in form and connected with interlobal cleft. Setae *d 4* 14 μ, setiform. Opisthosomal lobes slightly diverging posteriorly, with moderately developed lamella forming acute tip. Interlobal cleft with inverted U-shaped membrane. Setae *sh* 20 μ, spiculiform, setae *sR* 14 μ. Bases of setae *c 1* contiguous with pregenital apodeme.

The female of *Trouessartia fissispina* sp. n. differs from other

188

species in the combination of the following characteristics: presence of dorsal hysterosomal aperture, type of ornamentation of hysterosomal shield, long setae *l 1* and setiform setae *d 4* and form of lobar region.

Trouessartia megaplax sp. n. Fig. 48

Host and locality. – Blue-grey tanager, *Thraupis episcopus* (Linnaeus, 1766), local name blauwfoortje, Tawajariweg, 7.IX.1971 (1 ♂ holotype).
Deposition. – Rijksmuseum van Natuurlijke Historie, Leiden.

Male (holotype) (Fig. 48): Total length 583 μ, idiosomal length 528 μ, width 269 μ. Propodosomal shield 171 × 200 μ, very close to scapular shields: *sce-sce* 132 μ. Setae *l 1* 47 μ. Hysterosomal shield straight anteriorly, with deep lateral incisions. Terminal lamellae elongated, with smooth margins. Sternoventral sclerites are lacking. Genital organ 65 μ. Bases of setae *c 2* touching each other, these setae being situated distinctly posterior to setae *c 3*. Epimerites IVa slightly longer than adanal apodemes and with finger-like cranial projection. Adanal discs 12 × 12 μ, situated slightly before mid-distance between caudal part of genital organ and body terminus. Translobar apodeme developed.

The male of *T. megaplax* sp. n. differs from other species in the combination of the following characteristics: large propodosomal shield, long setae *l 1*, deep lateral incisions on hysterosomal shield and position of setae *c 2* and *c 3*.

Trouessartia minutipes Berlese, 1884

Pterocolus appendiculatus minutipes Berlese, 1884. A.M.S., fasc. *26*, nr. 4 (descr. ♂♀).
Trouessartia appendiculata var. *minutipes*, Canestrini & Kramer, 1899. Tierreich *7*: 121 (short diagn. ♂♀).

Host and locality. – Black-collared swallow, *Atticora melanoleuca* (Wied, 1820), Weg naar Zee, 10.XI.1971 (2 ♂♂, 1 ♀); White-lined tanager, *Tachyphonus rufus* (Boddaert, 1783), Weg naar Zee, 10.XI.1971 (1 ♂).

The species is known from various European and African Hirundinidae (GAUD & TILL, 1961). It is reported for the first time from South America.

Trouessartia aff. serrana Berla, 1959

Trouessartia serrana Berla, 1959. Bol. Mus. Nac., Rio de Janeiro, n.s., Zoologia nr. *209*: 7–9 (description ♂♀).

Host and locality. – Yellow-hooded blackbird, *Agelaius icterocephalus* (Linnaeus, 1766), local name geel-hede karoefowroe, Welgedacht, 22.VIII.1971 (2 ♀♀).

Our females correspond in main features to the characteristics of *T. serrana*, but differ from the figures accompanying the original description in some details: roundly triangular supranal concavity, long external spermathecal duct almost not tapering terminally, setae *sh* and *sR* with fine terminal hooklet. They may represent a closely related species. The host of *T. serrana* is also an icterid bird, *Ostinops decumanus* (BERLA, 1959). More material including males is needed for final conclusion.

Trouessartia unciseta sp. n.   Fig. 49-50

Host and locality. – Flame-crested tanager, *Tachyphonus cristatus* (Linnaeus, 1766), Weg naar Zee, 10.IX.1971 (♂ holotype and 3 ♀♀, 3 nymphs); White-lined tanager, *T. rufus* (Boddaert, 1783), Weg naar Zee, 10.IX.1971 (1 ♀, 1 nymph).

Deposition of type material. – Rijksmuseum van Natuurlijke Historie, Leiden; National Collection of Surinam, Paramaribo; University of Georgia, Athens; Institute of Parasitology, Prague.

Male (holotype) (Fig. 49): Total length 540 μ, idiosomal length 475 μ, width 246 μ. Propodosomal shield 131 × 142 μ, distant from scapular shields: *sce-sce* 97 μ. Setae *l 1* 41 μ. Hysterosomal shield anteriorly slightly concave, with deep lateral incisions. Terminal lamellae leaf-like, slightly overlapping, with smooth margins. Sternoventral sclerites are lacking. Setae *sh* 34 μ, with small terminal hooklet, setae *sh* 28 μ, strong, terminally curved. Genital organ 61 μ. Setae *c 2* on weakly sclerotized subgenital shield, with their bases touching, and slightly posterior to *c 3*. Epimerites IVa strongly reduced. Adanal discs 15 × 15 μ, situated slightly before mid-distance between caudal part of genital organ and body terminus. Translobar apodeme developed.

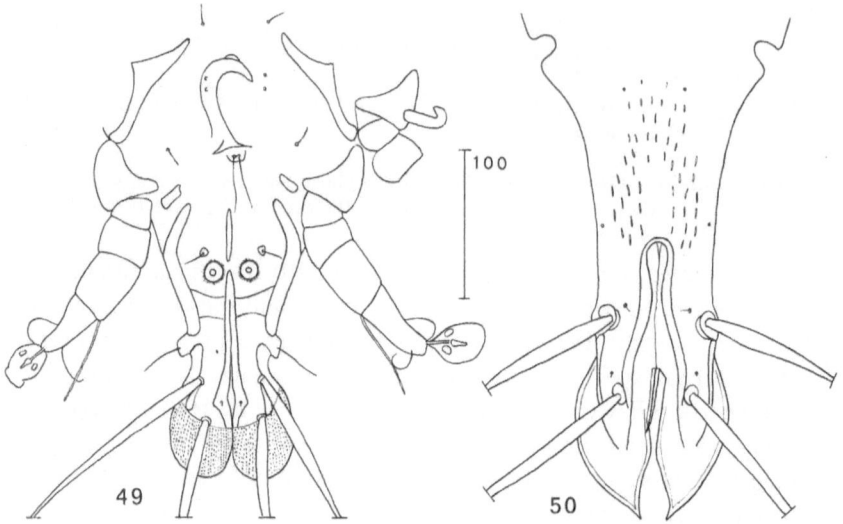

Fig. 49.   *Trouessartia unciseta* sp. n. – Male, body terminus, ventral view.

Fig. 50.   *Trouessartia unciseta* sp. n. – Female, body terminus, dorsal view.

F e m a l e (allotype) (Fig. 50): Total length 601 μ, idiosomal length 535 μ, width 268 μ. Propodosomal shield 133 × 158 μ, distant from scapular shields: *sce-sce* 112 μ. Setae *l 1* 36 μ, dilated and slightly bent. Hysterosomal shield anteriorly slightly concave, with deep lateral incisions and with elongated and fissiform lacunae in posterior half. Supranal concavity oval, connected with interlobal cleft. Setae *d 4* setiform, very fine. Opisthosomal lobes slightly diverging posteriorly, with bottom membrane and large lamellae terminating in acute tip. Spermathecal duct projecting into interlobal cleft, only slightly bent, tapering terminally to a needle-like tip. Setae *sh* 35 μ, strong, with fine terminal hooklet. Setae *sR* very strong, hook-like, 32 μ long. Bases of setae *c 1* contiguous with pregenital apodeme, *c 1-c 2* 17 μ.

Both sexes of *T. unciseta* sp. n. differ markedly from other species in the form of setae *sh* and *sR*.

Trouessartia spp.

Host and locality. – Pale-breasted thrush, *Turdus leucomelas* Vieillot, 1818, local name boontjedief, Welgedacht, 19.IX.1971 (1 ♂, 1 ♀, 8 nymphs); Bare-eyed thrush, *Turdus nudigenis* Lafresnaye, 1848, Tawajariweg, 7.IX.1971 (8 nymphs); Silver-beaked tanager, *Ramphocelus carbo* (Pallas, 1764), local name kieng, Tawajariweg, 7.IX.1971 (12 nymphs); Cinnamon-rumped foliage-gleaner, *Philydor pyrrhodes* (Cabanis, 1848), Tawajariweg, 9.IX.1971 (2 nymphs).

The male and female from *Turdus leucomelas* are in bad condition. The male differs from *Trouessartia mangaratibensis* Berla, 1959, known from *Turdus albicollis* from Brazil, in the form of lobal lamellae. The female of the latter species is unknown (BERLA, 1959). The nymphs from other hosts can only be identified on generic level.

Family ANALGIDAE Mégnin & Trouessart, 1883

Genus **Analges** Nitzsch, 1818

Subgenus *Analgopsis* Trouessart, 1919

Analges (Analgopsis) aff. corvinus Mégnin, 1877

Analges corvinus Mégnin, 1877. J. Anat. Physiol. *13*: 503 (descr. ♂♀).
Analges corvinus, Canestrini & Kramer, 1899. Tierreich *7*: 88 (short diagn. ♂♀).
Analges (Analgopsis) corvinus, Gaud & Mouchet, 1959. Ann. Par. hum. comp. *34*: 157 (body ventrally ♂).

Host and locality. – *Troglodytes aedon*, Welgedacht, 9.VIII.1971 (2 ♀♀, 3 nymphs, 1 larva); *T. aedon*, Welgedacht, 22.VIII.1971 (9 ♂♂, 19 ♀♀, 11 nymphs, 1 larva); *Agelaius icterocephalus*, Welgedacht, 22.VIII.1971 (1 ♂).

The genus *Analges* needs a revision. The records of *A. corvinus* from birds of other families than Corvidae may represent closely related species.

Analges sp.

Host and locality. – *Thraupis episcopus*, Tawajariweg, 7.IX.1971 (1 ♀).

Without males it is impossible to identify the members of this genus.

Genus **Diplaegidia** Hull, 1934

Diplaegidia columbigallinae Černý, 1975

Host and locality. – Ruddy ground dove, *Columbigallina talpacoti* (Temminck, 1811), local name stondovie, Weg naar Zee, 10.IX.1971 (1 ♂, 1 nymph).

Genus **Mesalgoides** Gaud & Atyeo, 1967

Subgenus *Mesalgoides* Gaud & Atyeo, 1967

Mesalgoides (Mesalgoides) elaeniae Černý, 1974

Host and locality. – *Elaenia flavogaster* (Thunberg, 1822), Weg naar Zee, 10.IX.1971 (1 ♀, 1 nymph).

Mesalgoides (Mesalgoides) furnarius Černý, 1974

Host and locality. – *Philydor pyrrhodes* (Cabanis, 1848), Tawajariweg, 9.IX. 1917 (1 ♀, 1 nymph).

Mesalgoides (Mesalgoides) koki Černý, 1974

Host and locality. – *Agelaius icterocephalus* (Linnaeus, 1766), Welgedacht, 22.VIII.1971 (17 ♂♂, 18 ♀♀, 15 nymphs, 3 larvae); *Troglodytes aedon* Vieillot, 1808, Welgedacht, 22.VIII.1971 (8 ♀♀, 1 nymph).

Mesalgoides (Mesalgoides) lukoschusi Černý, 1974

Host and locality. – *Thraupis episcopus* (Linnaeus, 1766), Tawajariweg, 17. IX.1971 (1 ♂, 2 ♀♀).

Mesalgoides (Mesalgoides) surinamensis Černý, 1974

Host and locality. – *Tachyphonus cristatus* (Linnaeus, 1766), Weg naar Zee, 10.IX.1971 (1 ♂, 1 nymph).

Mesalgoides (Mesalgoides) turdinus Černý, 1974

Host and locality. – *Turdus leucomelas* Vieillot, 1818, Welgedacht, 9.IX.1971 (2 ♂♂, 35 ♀♀, 10 nymphs); *Turdus nudigenis* Lafresnaye, 1848, Tawajariweg, 7.IX. 1971 (1 ♂, 1 ♀, 6 nymphs, 1 larva).

Subgenus *Chiasmalges* Gaud & Atyeo, 1967

Mesalgoides (Chiasmalges) sp.

Host and locality. – Black-headed parrot, *Pionites melanocephala* (Linnaeus, 1758), local name witbere prakiekie, Coronie, 27.VIII.1971 (1 ♀).

Chiasmalges was originally erected as a genus, but recently (GAUD, personal communication) it has been recognized as a subgenus of *Mesalgoides*. *Mesalgoides* (*Chiasmalges*) *polyplectrus* Gaud & Atyeo, 1967 from a Mexican parakeet *Aratinga holochlora*, is the only species included (GAUD & ATYEO, 1967). Our female represents another species which differs in the following characteristics: diverging branches of epimerites I shorter and situated more posteriorly, hysterosomal shield suboblong, deeply concave posteriorly, body terminus without triangular protuberance, setae *l 5* and *d 5* inserted ventrally. It is recommended to describe this species when the male will be available. It seems probable that the species of the subgenus *Chiasmalges* are confined to Psittaciformes.

Subgenus *Picalgoides* Černý, 1974

Mesalgoides (Picalgoides) capitonis Černý, 1974

Host and locality. – Black-spotted barbet, *Capito niger* (P. L. Statius Müller, 1776), Paramaribo, 15.VIII.1971 (14 ♂♂, 18 ♀♀, 7 nymphs, 1 larva).

Family FALCULIFERIDAE Oudemans, 1908

Genus Pterophagoides Gaud & Mouchet, 1959

Pterophagoides talpacoti Černý, 1975

Host and locality. – *Columbigallina talpacoti* (Temminck, 1811), Weg naar Zee, 10.IX.1971 (10 ♂♂, 12 ♀♀, 23 nymphs, 5 larvae).

Family PROCTOPHYLLODIDAE Mégnin & Trouessart, 1883

Subfamily Allodectinae Park & Atyeo, 1971

Genus **Allodectes** Gaud & Berla, 1963

Allodectes similis Černý, 1974

Host and locality. – Glittering-throated emerald, *Amazilia fimbriata* (Gmelin, 1788), Welgedacht, 27.VIII.1971 (9 ♂♂, 7 ♀♀).

Subfamily Proctophyllodinae Mégnin & Trouessart, 1883

Genus **Anisophyllodes** Atyeo, 1967

Anisophyllodes intermedius (Trouessart & Neumann, 1888)

Pterodectes intermedius Trouessart & Neumann, 1888. Bull. Sci. France Belgique *19*: 369–370 (descr. ♂♀).
Alloptes intermedius, Canestrini & Kramer, 1899. Tierreich *7*: 108 (short diagn. ♂♀).
Alloptes intermedius, Ateyo & Braasch, 1966, Bull. Univ. Nebraska St. Mus. *5*: 313–314 (no descr.).
Anisophyllodes intermedius, Ateyo, 1969. J. Georgia Entomol. Soc. *4*: 153–155 (redescr. ♂♀).

Host and locality. – *Elaenia flavogaster* (Thunberg, 1822), Weg naar Zee, 10.XI.1971 (1 ♂, 1 ♀, 6 nymphs).

The species is reported from *Elaenia martinica* and *Loxigilla noctis*. The latter host is a fringillid and must be considered a questionable record (ATYEO, 1969).

Genus **Proctophyllodes** Robin, 1877

Proctophyllodes atyeoi Černý, 1974

Host and locality. – *Agelaius icterocephalus* (Linnaeus, 1766), Welgedacht, 31.VIII.1971 (1 ♂, 1 ♀).

Proctophyllodes kratochvili Černý, 1974

Host and locality. – *Turdus leucomelas* Vieillot, 1818, Welgedacht, 19.IX.

1971 (1 ♂, 13 ♀♀, 5 nymphs); *T. nudigenis* Lafresnaye, 1848, Tawajariweg, 7.IX. 1971 (2 ♀♀).

Proctophyllodes parvilamellatus Černý, 1974

Host and locality. – *Philydor pyrrhodes* (Cabanis, 1848), Tawajariweg, 9.IX. 1971 (1 ♂, 1 nymph).

Proctophyllodes trisetosus Ewing & Stover, 1915

Proctophyllodes trisetosus Ewing & Stover, 1915. Entom. News *26*: 113–114 (descr. ♂♀).
Proctophyllodes trisetosus, Atyeo & Braasch, 1966. Bull. Univ. Nebraska St. Mus. *5*: 128–130 (redescr. ♂♀).

Host and locality. – Red-breasted blackbird, *Leistes militaris* (Linnaeus, 1758), local name reddie borstoe, Wageningen 23.IX.1971 (3 ♂♂, 7 ♀♀, 1 nymph).

This species is a parasite of icterid birds. *Leistes militaris*, *Sturnella magna* and *S. neglecta* are known as its hosts (ATYEO & BRAASCH, 1966).

Proctophyllodes sp.

Host and locality. – *Tachyphonus cristatus* (Linnaeus, 1766), Weg naar Zee, 10.IX.1971 (1 ♀).

The female does not correspond to any of the species known from Thraupidae (ATYEO & BRAASCH, 1966) and it is assumed to represent a new species. It is recommended to describe this species when the male will be available. The specimen is characterized by the propodosomal and hysterosomal shield covered with small lacunae, short (47 μ) lobar region, cleft in the form of an arch and distant setae *d 4* (46 μ) inserted on conjunctiva.

Subfamily Pterodectinae Park & Atyeo, 1971

Genus Pterodectes Robin, 1877

Pterodectes havliki Černý, 1974

Host and locality. – *Philydor pyrrhodes* (Cabanis, 1848), Tawajariweg, 9.IX. 1971 (8 ♂♂, 22 ♀♀, 3 nymphs, 3 larvae); *Ramphocelus carbo* (Pallas, 1764), Tawajariweg, 7.IX.1971 (5 ♂♂, 7 ♀♀); *Tachyphonus cristatus* (Linnaeus, 1766), Weg naar Zee,

10.IX.1971 (2 ♂♂, 3 ♀♀, 1 larva); *T. cristatus*, the same data (11 ♂♂, 33 ♀♀, 1 nymph, 2 larvae); *Atticora melanoleuca* (Wied, 1820), Weg naar Zee, 10.IX.1971 (1 ♀).

Pterodectes maculatus Černý, 1974

Host and locality. – *Agelaius icterocephalus* (Linnaeus, 1766), Welgedacht, 31.VIII.1971 (2 ♂♂).

Pterodectes reticulatus Černý, 1974

Host and locality. – *Elaenia flavogaster* (Thunberg, 1822), Weg naar Zee, 10.IX.1971 (2 ♀♀, 1 nymph, 3 larvae).

Pterodectes rutilus Robin, 1877

Proctophyllodes (Pterodectes) rutilus Robin, 1877. J. Anat. Physiol. *13*: 644 (descr. ♂♀).
Pterodectes rutilus, Canestrini & Kramer, 1899. Tierreich *7*: 124 (short diagn. ♂♀).
Pterodectes rhodesiensis Till, 1954. Moçambique Doc. Trim. *79*: 9P, 92–94 (descr. ♂♀).
Pterodectes rutilus, Gaud & Till, 1961. The Arthropod Parasites of Vertebrates in Africa South of the Sahara *1*: 255 (body ventrally ♂♀).
Pterodectes rutilus, Park & Atyeo, 1971. Bull. Univ. Nebraska St. Mus. *9*: 57 (body dorsally and ventrally ♂♀).

Host and locality. – *Atticora melanoleuca* (Wied, 1820), Weg naar Zee, 10.IX. 1971 (2 ♀♀, 2 nymphs).

The species is known from various European and African Hirundinidae. It is assumed to represent actually a species complex (PARK & ATYEO, 1971a).

Pterodectes storkani Černý, 1974

Host and locality. – *Ramphocelus carbo* (Pallas, 1764), Tawajariweg, 7.IX. 1971 (4 ♂♂, 7 ♀♀, 4 nymphs).

Pterodectes thraupicola Černý, 1974

Host and locality. – *Thraupis episcopus* (Linnaeus, 1766), Tawajariweg, 7.IX. 1971 (6 ♀♀).

Pterodectes troglodytis Černý, 1974

Host and locality. – *Troglodytes aedon* Vieillot, 1808, Paramaribo, 18.VII. 1971 (5 ♂♂, 4 ♀♀, 5 nymphs).

Pterodectes turdinus Berla, 1959

Pterodectus turdinus Berla, 1959. Bol. Mus. Nac. Rio de Janeiro, n.s., Zoologia nr.
  209: 11–14 (descr. ♂♀).

Host and locality. – *Turdus nudigenis* Lafresnaye, 1848, Tawajariweg, 7.IX.
1971 (30 ♂♂, 18 ♀♀, 15 nymphs, 3 larvae); *Turdus leucomelas* Vieillot, 1818, Welgedacht, 19.IX.1971 (10 ♀♀, 2 nymphs, 11 larvae); *Thraupis episcopus* (Linnaeus,
1766), Tawajariweg, 7.IX.1971 (1 ♂).

This species has been described from *Turdus rufiventris* from Brazil (BERLA, 1959).

Pterodectes sp.

Host and locality. – *Ramphocelus carbo* (Pallas, 1764), Tawajariweg, 7.IX.
1971 (1 ♂).

The specimen represents a new species which is not described here because of its
condition which does not allow to recognize all morphological structures. The male is
characterized by extremely large lacunae covering the whole surface of both dorsal
shields forming a reticuliform pattern.

Genus Trochilodectes Park & Atyeo, 1971

Trochilodectes brevicaulus Černý, 1974

Host and locality. – Little hermit, *Phaethornis longuemareus* (Lesson, 1832),
Zanderij, 6.IX.1971 (13 ♂♂, 10 ♀♀, 7 nymphs, 2 larvae).

Subfamily Rhamphocaulinae Park & Atyeo, 1971

Genus Rhamphocaulus Park & Atyeo, 1971

Rhamphocaulus sp.

Host and locality. – *Amazilia fimbriata* (Gmelin, 1788), Welgedacht, 27.VIII.
1971 (1 ♂, 3 ♀♀).

The specimens are not in good condition for description. They probably represent
a new species. The male has a penis without distal expansion, not reaching the level
of setae *c 2*. It is intermediate in certain characteristics between *R. vachoni* Park &
Atyeo, 1971 and *R. sinuatus* Park & Atyeo, 1971. Until now the genus *Amazilia* is

not represented among the hosts of the species of *Rhamphocaulus* (PARK & ATYEO, 1971b).

Family PTEROLICHIDAE Mégnin & Trouessart, 1883

Genus **Coraciacarus** Dubinin, 1956

Coraciacarus ani Černý, 1975

Host and locality. – Smooth-billed ani, *Crotophaga ani* Linnaeus, 1758, local name kawfoetoeboy, Paramaribo, 15.VIII.1971 (51 ♂♂, 56 ♀♀, 25 nymphs, 4 larvae); *C. ani*, Weg naar Zee, 17.IX.1971 (1 nymph).

Coraciacarus biemarginatus (Mégnin & Trouessart, 1884)

Pterolichus biemarginatus Mégnin & Trouessart, 1884. J. Microgr. *8*: 332 (descr. ♂♀).
Pterolichus (*Eupterolichus*) *biemarginatus*, Canestrini & Kramer, 1899. Tierreich *7*:
   47 (short diagn. ♂♀).

Host and locality. – *Capito niger* (P. L. Statius Müller, 1766), Paramaribo, 15.VIII.1971 (10 ♂♂, 17 ♀♀, 6 nymphs, 1 larva).

This mite parasitizes various birds of the family Capitonidae. *Capito niger* is known as host of this species (RADFORD, 1958).

Family PTERONYSSIDAE Dubinin, 1953

Genus **Pteronyssoides** Hull, 1931

Pteronyssoides atticorae Černý, 1975

Host and locality. – *Atticora melanoleuca* (Wied, 1820), Weg naar Zee, 10.IX. 1971 (11 ♂♂, 8 ♀♀, 10 nymphs, 3 larvae).

Genus **Pteronyssus** Robin, 1868

Pteronyssus hyalifer Černý, 1975

Host and locality. – *Capito niger* (P. L. Statius Müller, 1766), Paramaribo, 15.VIII.1971 (1 ♂).

Family XOLALGIDAE Gaud & Mouchet, 1959

Genus **Ingrassiella** Dubinin, 1949

Ingrassiella calcarata Černý, 1975

Host and locality. – *Turdus nudigenis* Lafresnaye, 1848, Tawajariweg, 7.IX. 1971 (1 ♂).

Genus **Xolalges** Trouessart, 1885

Xolalges tener Černý, 1975

Host and locality. – *Capito niger* (P. L. Statius Müller, 1766), Paramaribo, 15.VIII.1971 (11 ♂♂, 10 ♀♀, 4 nymphs); *Tachyphonus cristatus* (Linnaeus, 1766), Weg naar Zee, 10.IX.1971 (1 ♂).

The flame-crested tanager is probably an accidental host.

HOST–PARASITE RELATIONSHIPS

(*Probably accidental findings)

Psittaciformes

    PSITTACIDAE

        Pionites melanocephala (L.)   *Mesalgoides* sp.

Cuculiformes

    CUCULIDAE

        Crotophaga ani L.   *Coraciacarus ani* Černý, 1975

Columbiformes

    COLUMBIDAE

        Columbigallina talpacoti   *Diplaegidia columbigallinae* Černý, 1975
        (Temminck)   *Pterophagoides talpacoti* Černý, 1975

Apodiformes

TROCHILIDAE

Amazilia fimbriata (Gmelin)   *Allodectes similis* Černý, 1974
Rhamphocaulus sp.

Phaethornis longuemareus   *Trochilodectes brevicaulus* Černý, 1974
(Lesson)

Piciformes

CAPITONIDAE

Capito niger (Müller)   *Coraciacarus biemarginatus* (Mégnin & Trouessart, 1884)
Mesalgoides capitonis Černý, 1974
Pteronyssus hyalifer Černý, 1975
Xolalges tener Černý, 1975

Passeriformes

FURNARIIDAE

Philydor pyrrhodes (Cabanis)   *Mesalgoides furnarius* Černý, 1974
Proctophyllodes parvilamellatus Černý, 1974
Pterodectes havliki Černý, 1974
Trouessartia sp.

HIRUNDINIDAE

Atticora melanoleuca (Wied)   *Pterodectes havliki* Černý, 1974 *
Pteroctodectes rutilus Robin, 1877
Pteronyssoides atticorae Černý, 1975
Trouessartia appendiculata Berlese, 1884
Trouessartia minutipes Berlese, 1884

ICTERIDAE

Agelaius icterocephalus (L.)   *Analges* aff. *corvinus*
Mesalgoides koki Černý, 1974
Protoctophyllodes atyeoi Černý, 1974
Pterodectes maculatus Černý, 1974
Trouessartia aff. *serrana*

Leistes militaris (L.)   *Proctophyllodes trisetosus* Ewing & Stover, 1915

THRAUPIDAE

Ramphocelus carbo (Pallas)
Pterodectes havliki Černý, 1974
Pterodectes storkani Černý, 1974
Pterodectes sp.
Trouessartia sp.

Tachyphonus cristatus (L.)
Mesalgoides surinamensis Černý, 1974
Proctophyllodes sp.
Pterodectes havliki Černý, 1974
Trouessartia unciseta sp. n.
Xolalges tener Černý, 1975

Tachyphonus rufus (Boddaert)
Pterodectes havliki Černý, 1974
Trouessartia minutipes Berlese, 1884 *
Trouessartia unciseta sp. n.

Thraupis episcopus (L.)
Analges sp.
Mesalgoides lukoschusi Černý, 1974
Pterodectes thraupicola Černý, 1974
Pterodectes turdinus Berla, 1959 *
Trouessartia megaplax sp. n.

TROGLODYTIDAE

Troglodytes aedon Vieillot
Analges aff. *corvinus*
Mesalgoides koki Černý, 1974
Pterodectes troglodytis Černý, 1974
Trouessartia aedon sp. n.

TURDIDAE

Turdus leucomelas Vieillot
Mesalgoides turdinus Černý, 1974
Proctophyllodes kratochvili Černý, 1974
Pterodectes turdinus Berla, 1959
Trouessartia sp.

Turdus nudigenis Lafresnaye
Ingrassiella calcarata Černý, 1975
Mesalgoides turdinus Černý, 1974
Proctophyllodes kratochvili Černý, 1974
Pterodectes turdinus Berla, 1959
Trouessartia sp.

TYRANNIDAE

Elaenia flavogaster (Thunberg)
Anisophyllodes intermedius (Trouessart & Neumann, 1888)
Mesalgoides elaeniae Černý, 1974
Pterodectes reticulatus Černý, 1974
Trouessartia fissispina sp. n.

From our results, which are summarized above, the following facts may be emphasized. First of all, the feather mite fauna found on birds from Surinam is very rich. On 18 host species listed in 6 orders and 12 families at least 44 analgoid species have been collected. From most passeriform birds 4–5 feather mite species have been recorded. Moreover, in most instances these 4–5 analgoid species were found on a single host specimen. These species are: *Philydor pyrrhodes, Atticora melanoleuca, Ramphocelus carbo, Tachyphonus cristatus, Thraupis episcopus, Turdus leucomelas, Turdus nudigenis* and *Elaenia flavogaster*.

Secondly, in some bird species more than one member of a certain feather mite genus occurs, e.g. 3 *Pterodectes* species on *Ramphocelus carbo*, 2 *Trouessartia* species on *Atticora melanoleuca*. In some cases of double infestation one host-parasite association seems to be merely accidental, e.g. *Atticora melanoleuca* and *Pterodectes havliki, Tachyphonus rufus* and *Trouessartia minutipes*.

Some conclusions about the host specificity of the representatives of individual genera, such as *Mesalgoides, Pterodectes* and partly also *Trouessartia*, can only be made when more species have been collected. The *Mesalgoides* species are mostly monoxenous, only *M. turdinus* has monogeneric distribution; the situation with *M. koki* is not clear. It seems that also the Old World species of *Mesalgoides* from passeriform birds will show more diversity than is believed nowadays. Also most of *Pterodectes* species are probably monoxenous. *P. turdinus* has monogeneric distribution, *P. rutilus* a monofamilial one. Only *P. havliki* seems to have monoordinal distribution. In the genus *Trouessartia* both monoxenous (*T. aedon, T. fissispina, T. megaplax*) and oligoxenous species (*T. unciseta, T. appendiculata, T. minutipes*) have been found. Further research may change these categories, especially in monoxenous species.

REFERENCES

ATYEO, W. T., 1969. Redescription of Anisophyllodes intermedius (Trouessart and Neumann), 1888, new combination (Acarina: Proctophyllodidae). *J. Georgia Entom. Soc. 4*: 153–155.

ATYEO, W. T. & BRAASCH, N. L., 1966. The feather mite genus Proctophyllodes (Sarcoptiformes: Proctophyllodidae). *Bull. Univ. Nebraska State Mus. 5*: 1–354.

BERLA, H. F., 1959. Analgesoidea neotropicais. IV. – Sôbre algunas espécies novas ou pouco conhecidas de acarinos plumícolas. *Bol. Mus. Nac. Rio de Janeiro* (n. s.) *Zoologia 209*: 1–17.

ČERNÝ, V., 1974a. Parasitic mites of Surinam XIX. Seven new species of Mesalgoides (Analgoidea, Analgidae). *Folia parasit.* (Praha) *21*: in press.

ČERNÝ, V., 1974b. Parasitic mites of Surinam XXXI. New species of Proctophyllodidae (Sarcoptiformes, Analgoidea). *Folia parasit. 21*: in press.

ČERNÝ, V., 1975. Parasitic mites of Surinam XXXII. New species of feather mites (Sarcoptiformes, Analgoidea). *Folia parasit. 22*: in press.

GAUD, J. & ATYEO, W. T., 1967. Genres nouveaux de la famille des Analgidae, Trouessart et Mégnin. *Acarologia 9*: 447–464.

GAUD, J. & TILL, W., 1961. Genus Trouessartia Canestrini. In: ZUMPT, F., *The arthropod parasites of vertebrates in Africa south of the Sahara, I.* Publ. South Afr. Ins. Med. Res. *50*: 259–263.

PARK, C. K. & ATYEO, W. T., 1971a. A generic revision of the Pterodectinae, a new subfamily of feather mites (Sarcoptiformes: Analgoidea). *Bull. Univ. Nebraska State Mus. 9*: 39–88.

PARK, C. K. & ATYEO, W. T., 1971b. A new subfamily and genus of feather mites from hummingbirds (Acarina: Proctophyllodidae). *Fla. Entomol. 54*: 221–229.

RADFORD, C. D., 1958. The host-parasite relationships of the feather mites (Acarina: Analgesoidea). *Rev. Brasil. Ent. 8*: 107–170.